A PRACTICAL INTRODUCTION TO CERAMIC SCIENCE & INDUSTRY

CERAMICS MADE SIMPLE

(A Behaviour-Based Guide to Ceramic Materials, Processing, and Failure)

A PRACTICAL INTRODUCTION TO CERAMIC SCIENCE & INDUSTRY

CERAMICS MADE SIMPLE

(A Behaviour-Based Guide to Ceramic Materials, Processing, and Failure)

SANKAMEESWARAN S

notionpress.com

INDIA • SINGAPORE • MALAYSIA

ISBN
Paperback 979-8-90431-097-4
Hardcase 979-8-90431-098-1

Dedicated

To those who work with ***fire***, ***pressure***, and ***structure*** —
— and choose discipline over assumption.

Disclaimer

This book is intended to provide general information on ceramic materials, processing, and failure mechanisms based on established engineering principles and industrial experience.

The information is provided for educational and reference purposes only. Although every effort has been made to ensure accuracy, the author and publisher make no warranties regarding the completeness or applicability of the material for specific industrial situations.

Ceramic systems operate under a wide range of materials, processing conditions, and service environments. Readers should apply appropriate professional judgment, testing, and verification when implementing any concepts or practices described in this book.

The author and publisher shall not be liable for any loss, damage, or operational consequences arising from the use or interpretation of the information presented.

CERAMICS MADE SIMPLE is a structured guide to understanding how ceramic materials truly behave in industrial conditions. Written for engineers, plant professionals, designers, and students, this book moves beyond surface-level explanations and organizes ceramic science into a clear system of cause and effect.

From raw materials and microstructure to thermal expansion, *firing*, *glaze fit*, *refractories*, and *failure diagnosis*, it connects processing decisions to real-world performance. Complex topics such as brittleness, residual stress, thermal shock, creep, and flaw-controlled fracture are explained in disciplined, practical language—without unnecessary mathematics.

Rather than treating defects as isolated problems, the book presents ceramics as structured systems governed by physical laws. It builds decision intelligence for production, quality control, installation, and troubleshooting.

Ceramics Made Simple does not make ceramics shallow.
It makes them understandable, predictable, and engineerable.

Preface

Ceramics are everywhere.
Tiles beneath our feet.
Sanitaryware in daily use.
Refractories protecting furnaces.
Insulators in power systems.
Advanced components in aerospace and electronics.

Despite their widespread use, the behaviour of ceramics is frequently misunderstood. Defects are often treated as surface problems. Failures are often corrected at the symptom level. Process adjustments are often made without structural diagnosis of the problem.

Ceramics do not fail randomly. They respond to structure, stress, temperature, and constraint.

Understanding these relationships is the foundation of reliable ceramic engineering.

This book takes a behaviour-based approach. It begins with the fundamental laws that govern ceramic materials. It then moves through processing, microstructure formation, thermal behaviour, mechanical response, glaze interaction, and refractory performance. Finally, it connects these principles to failure analysis and engineering decisions.

The emphasis throughout is on cause and effect.

Every chapter follows a disciplined structure:

- Behaviour explained clearly
- Industrial relevance established
- Failure logic connected
- Decision insight provided

The goal is not academic depth detached from practice. The goal is structural clarity that engineers can apply in real manufacturing environments.

This book is written for production engineers, quality engineers, plant managers, refractory specialists, and students of ceramic engineering who want to understand not only what happens, but why it happens.

Ceramic materials do not behave randomly. Their behaviour follows identifiable physical mechanisms controlled by processing, structure, and operating conditions. This book aims to make those mechanisms visible so that ceramic performance and failure can be understood logically rather than treated as unpredictable events.

The Ceramic Behaviour System (CBS), explained fully in Chapter 1, provides the framework used in this book to connect processing, structure, and performance.

Ceramics Made Simple does not reduce scientific depth.

It removes confusion by revealing the structural logic behind ceramic behaviour.

Table of Contents

List of Figures

All figures in this book are schematic illustrations prepared by the author for explanatory purposes unless otherwise stated.

List of Tables

Who This Book is For

This book is written for people who work with ceramics in real conditions — not ideal ones.

It is for professionals who must make decisions, prevent defects, and understand failure.

Production Engineers, who control forming, drying, firing, and cooling, and need to understand how each decision affects structure.

Quality Engineers, who diagnose cracks, warpage, glaze defects, and refractory degradation.

Plant Managers, who balance performance, cost, yield, and reliability.

Design Engineers, who apply ceramics under thermal and mechanical load.

Refractory Engineers, who manage lining systems exposed to high temperature, stress, and chemical attack.

Students of Ceramic Engineering, who want conceptual clarity before specialization.

How This Book is Written

This book assumes basic familiarity with materials and manufacturing processes.

It does not require advanced mathematics. It explains Ceramic Behaviour System through structure, stress, temperature, and process logic.

Its purpose is simpler: to provide a structured understanding of how ceramics behave in real manufacturing environments.

Ceramics fail when structure and stress are misunderstood.

For this reason, the book focuses on a few essential ideas:

- Structural thinking
- Thermal intelligence
- Mechanical truth
- Process discipline
- Failure diagnosis

Short industrial examples are also included.

The goal is clarity that engineers can use in practice.

How to Use This Book

The book is organized in progressive layers.

Part I–III explain the fundamental behaviour of ceramics.
Part IV–VIII examine processing, firing, glaze behaviour, and common defects.
Part IX addresses refractories operating under extreme temperature conditions.
Part X integrates the principles into a structured approach to failure diagnosis.

Readers new to ceramics may begin at Part I and proceed sequentially.

Experienced engineers may consult chapters relevant to a specific process stage or failure mode.

Figures provide conceptual models of behaviour.

Tables summarize important comparisons and decision points.

This book is not intended to be read only once.

It is designed to serve as a reference framework when diagnosing problems in manufacturing or making engineering decisions.

Terminology Discipline

How Terms Are Used in This Book

Ceramic engineering requires precision in language. The following definitions explain how key terms are used throughout this book.

Strength: Unless otherwise specified, strength refers to tensile-dominated fracture strength. Ceramics are weak in tension and strong in compression; tensile behaviour governs failure.

Stress: Internal force per unit area within a material. It is not the same as applied load. Load creates stress; geometry distributes it.

Strain: Deformation per unit length. In ceramics, strain remains elastic until fracture.

Density: Unless specified otherwise, density refers to bulk density. Apparent and True density are used only when explicitly mentioned.

Porosity: Total void fraction within the structure. Open and Closed porosity are distinguished where permeability is relevant.

Microstructure: The internal arrangement of grains, phases, pores, and residual stresses formed during processing and firing.

Residual Stress: Stress locked into the material without external load, typically generated by thermal gradients or phase transformations.

Thermal Shock: Fracture caused by rapid temperature change creating internal tensile stress.

Creep: Time-dependent deformation under sustained load at elevated temperature.

Failure: Structural or functional breakdown caused by stress exceeding material capacity, including cracking, spalling, deformation, or chemical degradation.

Compatibility: Alignment of thermal expansion, mechanical behaviour, and chemical stability between interacting materials.

Why Terminology Discipline Matters:

This book treats ceramics as systems governed by structure and stress. When terms are used consistently:

- Diagnosis becomes clearer.
- Decisions become more precise.
- Communication improves across teams.

Precision in language supports precision in engineering.

What This Book Does NOT Cover

This book is intentionally focused. It does not attempt to cover every branch of ceramic science.

It does not provide detailed mathematical derivations of fracture mechanics.
It does not explore advanced ceramic composites in depth.
It does not cover electro ceramics or semiconductor ceramics extensively.
It does not replace plant trials or material certification testing.
It does not promise perfect prediction of failure.

Ceramics are statistical materials. Variability cannot be fully eliminated.

This book is designed to provide:

- Structural clarity
- Practical industrial understanding
- Decision-oriented reasoning
- Failure diagnosis logic

It builds understanding from raw materials to failure.
It does not attempt to be an encyclopaedia.
Its strength is coherence.

About The Author

Sankameeswaran S works at the intersection of ceramic science and industrial practice, with a focus on bringing structural clarity to materials behaviour. His work examines the mechanisms that govern ceramic performance, particularly failure behaviour, process–property relationships, and decision frameworks within ceramic systems.

With a strong emphasis on industrial application, his approach connects theoretical understanding with plant-level implementation.

In addition to his industrial work, Sankameeswaran S is also a LinkedIn educator, where he shares practical insights on ceramic materials, processing behaviour, and failure mechanisms. Through these contributions, he translates complex ceramic science into clear, behaviour-based understanding for engineers, plant professionals, and students across the global ceramic community.

Ceramics Made Simple reflects a structured consolidation of this perspective, presenting ceramic behaviour through clear mechanisms rather than fragmented theory.

Readers' Feedback

Understanding ceramic materials improves when plant experience and theory meet. Readers are welcome to share observations, corrections, or field insights related to the topics discussed in this book. Constructive feedback may contribute to improvements in future editions. Email: ssankam@hotmail.com

Ceramic Behaviour System — The Framework of This Book

Ceramic materials do not behave randomly. Their performance follows a structural chain linking raw materials, processing, microstructure, and service conditions.

This book uses the **Ceramic Behaviour System (CBS)** to explain that chain. The CBS connects raw materials, manufacturing processes, microstructure formation, and operating conditions to show how ceramic performance and failure develop. The complete explanation of this system is presented in Chapter 1 — The Ten Laws of Ceramic Behaviour.

The chapters of this book follow this structure progressively—from fundamental material behaviour to processing stages, thermal and mechanical response, and finally failure analysis.

Part I — How Ceramics Think (Ch. 1–5)
Foundations of ceramic behaviour: ten laws of ceramic behaviour, material identity, brittleness, flaw sensitivity, and microstructure.

Part II — Thermal Intelligence (Ch. 6–9)
Thermal response of ceramics: residual stresses, thermal expansion, thermal shock, thermal fatigue, and heat flow.

Part III — Mechanical Truth (Ch. 10–14)
Strength in brittle materials: the meaning of strength, size effect, time-dependent behaviour, and the difference between laboratory testing and real service conditions.

Part IV — Density & Porosity (Ch. 15–17)
The role of density and porosity in ceramic performance: density development, porosity formation, and the relationship between density and strength.

Part V — Raw Materials Intelligence (Ch. 18–20)
Understanding behaviour of raw materials: clay plasticity, quartz–feldspar balance, and the influence of particle size distribution, purity, and body design.

Part VI — Forming & Green Stage (Ch. 21–23)
Shaping ceramic bodies before firing: rheology, forming methods, and the origin of green-stage defects.

Part VII — Drying & Firing (Ch. 24–26)
Transformations during thermal processing: drying physics, sintering and vitrification, along with defects associated with heating and cooling.

Part VIII — Glaze (Ch. 27–28)
Glaze–body interactions: glaze fit and identification of common surface defects.

Part IX — Refractories (Ch. 29)
Ceramic performance under extreme temperature conditions: refractories and their common failure mechanisms.

Part X — Failure Intelligence (Ch. 30)
Practical failure analysis in ceramics: diagnosis, decision-making, and prevention.

The Central Thesis of This Book

Ceramics are often described as brittle, complex, or unpredictable. In reality, ceramic behaviour is governed by a disciplined structure.

A ceramic does not fail suddenly. The visible crack is only the final step of a structural sequence that began much earlier.

For this reason, ceramic engineering is not the memorization of isolated topics. It is the ability to read the system.

This book therefore treats ceramics as a connected behaviour framework — not as separate chapters on strength, expansion, firing, glaze, refractories, and defects.

Across tiles, sanitaryware, kiln furniture, and refractory linings, the governing logic is the same. Ceramic behaviour follows a connected structural relationship between materials, processing, structure and service conditions.

The behaviour of all ceramic materials can be understood through a small set of structural principles. These principles are expressed here as the Ten Laws of Ceramic Behaviour for the foundation of all ceramic systems and are presented in Chapter 1. They are referenced throughout the book as a framework for analysing ceramic behaviour and failure.

Part I — How Ceramics Think

Understanding the Behaviour of Ceramic Materials

This section establishes the fundamental behaviour of ceramic materials. Before examining processing or industrial defects, it is necessary to understand how ceramics respond to stress, temperature, and structure.

The chapters in this part explain the physical principles that govern ceramic behaviour and form the foundation for the Ceramic Behaviour System introduced in this book.

Ceramics differ fundamentally from metals and polymers in their bonding, deformation behaviour, and fracture response.

Fundamental Drivers of Ceramic Behaviour:

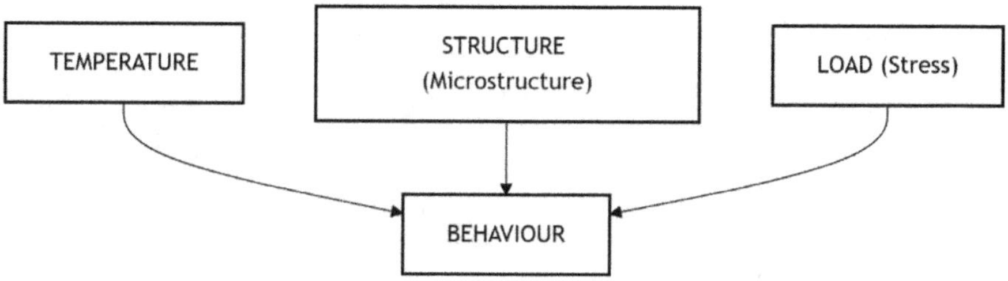

Ceramic behaviour is governed by the interaction of three primary factors: temperature, structure (microstructure), and applied load (stress).

Changes in any one of these conditions alter how the material performs or fails. Understanding this relationship is essential for predicting ceramic behaviour in processing and service environments.

This part introduces the foundational principles governing Ceramic Behaviour System. Concepts such as brittleness, flaw sensitivity, stress concentration, and microstructure are explained as parts of a single system. These principles form the intellectual framework used throughout the rest of the book.

Chapter 1

The Ten Laws of Ceramic Behaviour

Ceramic performance is governed by structure, not by chance.

Ceramics do not fail suddenly. They reveal the truth of their structure. When a ceramic tile cracks, the crack is not the beginning of the problem. It is the final step in a chain of events that may have started during mixing, pressing, drying, firing, cooling, or installation.

Ceramics do not break randomly. They respond to how they were made and how they are used. When this response is understood, failure becomes predictable. When it is ignored, failure becomes inevitable.

Unlike metals or polymers, ceramics cannot bend to relieve stress or stretch to hide mistakes. When the internal structure is balanced and service conditions are controlled, ceramics perform reliably. When imbalance exists, cracks appear.

Ceramic performance is governed by three interacting elements:

Structure — Stress — Environment

When these remain balanced, ceramics perform reliably. When imbalance accumulates, failure becomes inevitable.

CBS — Ceramic Behaviour System

Ceramic performance emerges from a chain beginning with raw materials and processing. Ceramic behaviour follows the **processing–structure–performance chain**.

This behavioural chain explains why ceramic failures rarely originate where they first appear. A crack observed in service may have its origin in microstructure, which in turn may trace back to processing conditions or raw material characteristics.

The Ceramic Behaviour System (CBS) therefore provides a framework for tracing failures backward through the chain of cause and effect.

All ceramic defects discussed in this book can be analysed through this framework. Engineering decisions become clearer when each stage of the chain is understood and controlled.

Failure rarely begins in service.
It begins in structure.

In real systems, ceramic components rarely expand freely. Installation methods, anchors, frames, glaze layers, or bonding interfaces often restrict movement. When thermal expansion is restrained, internal stress increases significantly.

Constraint amplifies thermal stress: when thermal expansion or contraction is restricted, even small temperature changes can generate significant tensile stress.

Chemistry also interacts with structure. Chemical environments such as slag, corrosion, or moisture can gradually weaken grain boundaries, alter phases, and reduce strength over time.

Understanding ceramics therefore requires reading the behaviour chain in reverse.

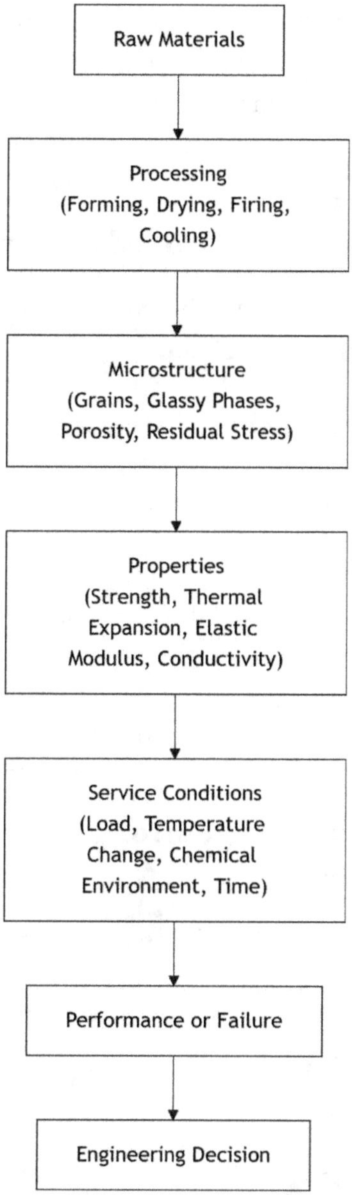

Figure 1.1 — Ceramic Behaviour System (CBS)
Relationship between processing, microstructure, properties, and service performance in ceramic materials.

If the structure is uneven, performance will also become uneven. When we see a crack, we are observing the final stage of a long sequence.

Ceramics are often studied through individual properties such as strength, thermal expansion, porosity, phase composition, or fracture behaviour. In real systems, these properties do not operate independently. They arise from an interconnected chain that begins with raw materials and ends with service performance.

First Law — Strength is controlled by flaws

Ceramic strength is not determined by average quality. It is controlled by the largest defect within the stressed region.

That defect may be a pore, a microcrack, a weak interface, or a density variation.

When tensile stress acts on the material, stress concentrates at the tip of that defect. Once the local stress exceeds a critical limit, crack growth begins.

Two tiles from the same batch may fail at different loads because their defect sizes are different.

In ceramics, small defects can have large consequences.

Second Law — Ceramics are strong in compression but vulnerable in tension

Compression tends to close cracks, while tension opens them.

Most ceramic failures originate in regions where tensile stress exists.

Tensile stress may arise from:

- cooling after firing
- bending loads
- glaze mismatch
- restrained thermal expansion during installation

The problem is not load alone.
The problem is tensile stress acting on flaws.

Third Law — Ceramics do not yield before fracture

Metals can bend and redistribute stress. Ceramics cannot.

Ceramics remain elastic until fracture occurs. There is almost no visible warning before failure.

When the critical limit is reached, fracture occurs suddenly.

Because ceramics lack plastic stress redistribution, design must always include sufficient safety margin.

Fourth Law — Processing defines microstructure

Ceramics remember how they were made.

If pressing is uneven, density varies across the body.
If drying is too fast, internal stresses develop.
If firing is uneven, phases form unevenly.
If cooling is too rapid, residual stress remains inside the structure.

The internal structure of ceramics is therefore built step by step during manufacturing.

Fifth Law — Microstructure defines properties

Strength, thermal expansion, durability, and heat resistance all depend on internal structure.

Grain size, porosity, phase distribution, and residual stress influence behaviour.

Two ceramics made from the same raw materials may behave differently if their microstructures differ.

Composition provides potential. Structure determines performance.

Sixth Law — Temperature gradients create stress

Heat alone does not cause failure. Uneven temperature does.

When different regions of a ceramic body expand or contract at different rates, internal stress develops.

Fast cooling can create tensile stress at the surface.
Temperature differences between the surface and interior generate internal stress fields.

Thermal shock is therefore a consequence of temperature gradients under restraint.

Seventh Law — Constraint amplifies thermal stress

Ceramic components rarely expand freely in real systems.

- Movement may be restricted by
- supports
- anchors
- frames
- glaze layers
- installation methods

When expansion or contraction is restrained, internal stress increases significantly.

Constraint therefore multiplies the stress created by temperature change.

Eighth Law — Time modifies behaviour

Ceramics do not respond only at one moment.

Under sustained load, cracks can grow slowly.
At high temperature, materials may creep gradually.
Repeated thermal cycling can extend microcracks.

A product that passes a short laboratory test may still fail after long service.

Time must always be considered in ceramic reliability.

Ninth Law — Chemistry interacts with structure

Chemical environments can weaken ceramic systems gradually.

Slag attack, corrosion, moisture interaction, and reactions at grain boundaries may degrade structure over time.

Chemical compatibility therefore becomes critical for long-term ceramic performance.

Tenth Law — Systems fail, not materials alone

Ceramic components operate inside systems.

These systems include:

- geometry
- supports
- anchors
- installation practices
- temperature fields

A well-manufactured ceramic body can fail if the surrounding system is poorly designed.

Tiles may tent when expansion joints are missing.
Refractories may crack when anchors are improperly positioned.

Movement that is blocked becomes stress.

When a failure appears, the important question is not only:

"What broke?"

The more important question is:

"Where in the behaviour chain did the imbalance begin?"

The Ceramic Operating System Model

Ceramic performance is governed by the interaction of material structure, geometry, constraint, temperature and time.

These variables determine the internal stress distribution that ultimately leads to stable performance or failure.

This model reminds us that ceramic performance depends on both the material and the system in which it operates.

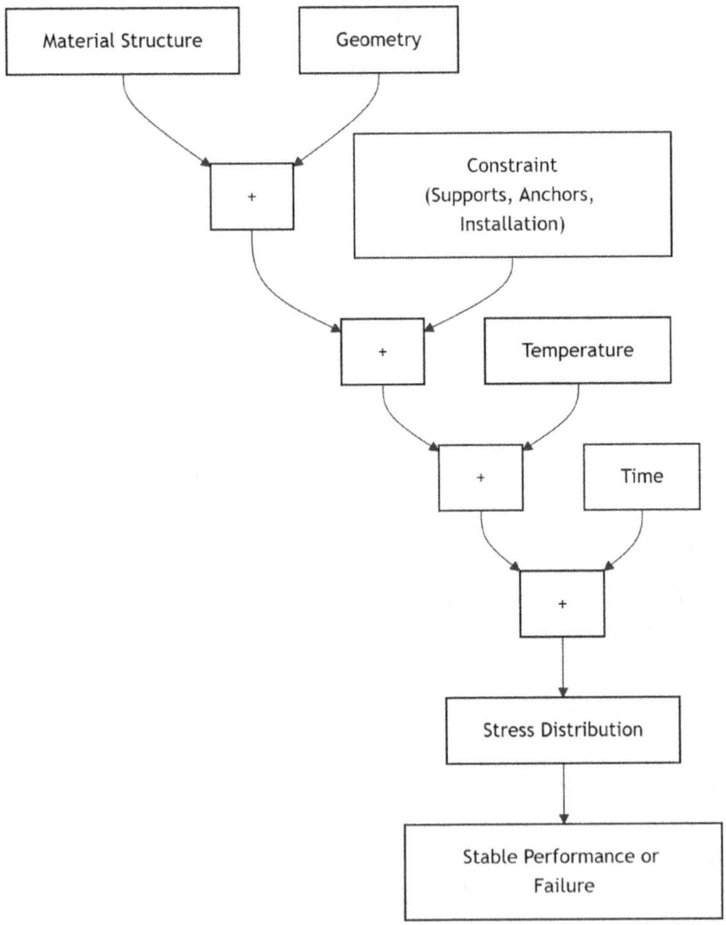

Figure 1.2 — CBS Operating Model
Ceramic performance depends on the interaction of material structure, geometry, constraint, temperature, and time.

Summary of the Ten Laws of Ceramic Behaviour

Table 1.1 — The Ten Laws of Ceramic Behaviour and Their Engineering Meaning

Law	Core Statement	Engineering Meaning
1	Strength is controlled by the largest flaw	Defect size governs fracture strength
2	Ceramics are strong in compression and vulnerable in tension	Control tensile stress zones
3	Ceramics do not yield before fracture	Design for brittle behaviour
4	Processing defines microstructure	Failures often originate during manufacturing
5	Microstructure defines properties	Internal structure determines real performance
6	Temperature gradients create stress	Manage thermal gradients and cooling rates
7	Constraint amplifies thermal stress	Allow expansion and avoid restrained movement
8	Time modifies behaviour	Evaluate creep, fatigue, and long-term reliability
9	Chemistry interacts with structure	Ensure chemical compatibility in service environments
10	Systems fail, not materials alone	Diagnose the entire operating system

Ceramic materials are not fragile. They are precise materials.

They behave according to structure. When structure is correct and conditions are controlled, performance is reliable. When structure is ignored, failure follows.

The chapters ahead apply these laws to thermal expansion, mechanical behaviour, porosity, firing, glaze fit, and refractory systems.

Ceramics are predictable once their behaviour is understood.

Decision Insight:

- Ceramic performance always originates in structure.
- Failures must be traced backward through the Ceramic Behaviour System chain.
- Correct structure produces reliable behaviour.

Chapter 2

What Ceramics Are — And What They Are Not

Ceramic materials are rigid, flaw-controlled materials that do not tolerate structural imbalance.

In the previous chapter we established a central principle: ceramic performance is governed by structure. Failures do not appear randomly; they emerge from the interaction of structure, stress, and environment. Before examining specific behaviours such as thermal shock, porosity, or glaze fit, we must clearly understand the nature of the material itself.

Many ceramic failures originate from misunderstanding how ceramic materials behave under stress, temperature, and constraint. A designer may assume that the material will bend to relieve stress. An installer may assume that restrained movement will be tolerated. A user may assume that impact energy will be absorbed. When these assumptions are incorrect, cracks appear.

Understanding ceramics therefore begins with a simple but critical distinction: what ceramics are — and what they are not.

Formally, ceramics are inorganic, non-metallic materials that are shaped and hardened by heat. This definition describes how ceramics are produced, but it does not fully explain how they behave. Behaviour is determined by the internal structure of the material.

Most ceramics are built from strong ionic and covalent bonds. These bonds hold atoms firmly in fixed positions within the crystal lattice. Because the bonds are strong and directional, the atomic structure becomes highly stable and resistant to movement.

This bonding structure gives ceramics several important advantages. They are stiff and resist deformation. They maintain their shape even at high temperatures. They often exhibit excellent resistance to chemical attack. Unlike many metals and polymers, they do not soften easily when exposed to heat.

However, the same bonding structure that gives ceramics their strength also limits their flexibility. Atomic layers cannot easily slide past each other as they do in metals. When stress increases beyond the elastic limit, the structure cannot adjust through plastic deformation. Instead, cracks initiate and propagate through the material.

This balance between strength and limitation defines the identity of ceramics. They are strong, stable and chemically resistant materials, but they are also highly sensitive to flaws and tensile stress.

To understand this behaviour more clearly, ceramics must be compared with the other major classes of engineering materials.

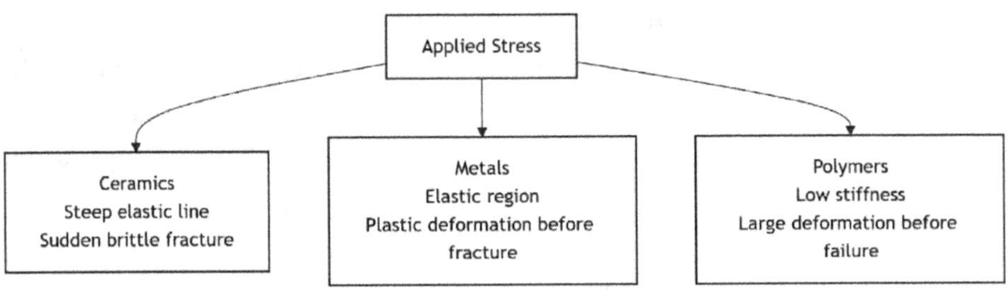

Figure 2.1 — Behaviour Comparison: Ceramics vs Metals vs Polymers
Different bonding structures produce distinct deformation and fracture behaviour.

This comparison highlights a fundamental behavioural difference.

Ceramics exhibit a steep elastic response and fracture suddenly with almost no plastic deformation.

Metals first deform elastically and then plastically, allowing noticeable deformation before fracture.

Polymers have lower stiffness and can undergo large deformation before failure because their molecular chains can move and rearrange.

Metals are able to bend. Under increasing load, they first respond elastically and then enter a plastic region where permanent deformation occurs. This plastic deformation absorbs energy, redistributes stress and often provides visible warning before fracture.

Polymers can stretch even further. Their long molecular chains are flexible and capable of rearranging under stress, allowing them to absorb impact energy through deformation.

Ceramics behave differently. Their stress–strain response is almost entirely linear. They remain elastic until the critical limit is reached, and then fracture occurs. Because there is almost no plastic deformation before failure, fracture appears sudden.

This behaviour explains why ceramics often appear both strong and fragile. They are strong because their atomic bonds create high stiffness and compressive strength. They appear fragile because they cannot relieve stress through deformation.

In compression, ceramics can carry extremely high loads. This is why bricks support buildings and refractory linings contain molten metals in furnaces. However, the situation changes when tensile stress is present.

Compressive stress tends to close cracks, while tensile stress opens them. Most real ceramic failures involve tensile stress somewhere in the component, even when the overall load appears compressive.

Tensile stress may develop in several ways. It can arise during cooling after firing, from bending during service, from mismatch between glaze and body, or from restrained thermal expansion after installation. When tensile stress interacts with a flaw, fracture begins.

Ceramics are also highly sensitive to defects because stress cannot redistribute through plastic deformation. A small pore, microcrack or density variation can concentrate stress and initiate failure. As a result, internal structure becomes critically important. Two ceramics made from the same raw materials may behave differently if their porosity, grain size or residual stress patterns differ.

Ceramics also behave differently from other materials at high temperature. Many metals soften when heated, and polymers may melt or degrade. Many ceramics, however, remain structurally stable in kilns and furnaces.

High temperature stability, however, does not mean resistance to temperature change. Rapid heating or cooling can create temperature gradients within the material. These gradients generate internal stresses that may lead to cracking.

For this reason, ceramics must not be treated as forgiving materials. They rarely provide visible warning before fracture. When they are expected to behave like metals or polymers, design errors occur.

Correct material selection therefore begins with behavioural clarity.

Summary Comparison

Table 2.1 — Ceramics vs Metals vs Polymers

Property	**Ceramics**	**Metals**	**Polymers**
Bonding Type	Ionic / Covalent	Metallic	Molecular Chains
Plastic Deformation	Very low	High	High
Failure Mode	Sudden fracture	Bending then fracture	Stretching / tearing
High Temperature Stability	Very high	Moderate	Low
Flaw Sensitivity	High	Medium	Low
Energy Absorption	Low	High	Medium

This comparison is not theoretical. It directly guides engineering decisions.

If an application requires high temperature stability and chemical resistance, ceramics may be the best choice. If an application requires bending tolerance, impact resistance or energy absorption, geometry and support conditions must be carefully designed, or another material may be more suitable.

Ceramic materials follow structural rules closely. When used within those rules, they perform extremely well. When those rules are ignored, failure follows.

The chapters ahead will explore how stress, temperature and processing interact with this behaviour. But the foundation is now clear.

Ceramics do not behave like metals.
They do not behave like polymers.

They behave according to their structure.

Decision Insight:

- Ceramics do not bend to relieve stress.
- Design must prevent tensile stress zones.
- Material selection must match loading mode.

Chapter 3

Brittleness, Elasticity and Fracture

Ceramics store elastic energy during loading and release it suddenly when fracture occurs.

Ceramics resist strongly, but they do not bend to survive. When a steel rod is pulled, it stretches and eventually bends before breaking. The deformation becomes visible as the load increases. A ceramic rod behaves differently. It stretches only slightly—almost invisibly—and then fractures. There are no bending stage and no visible warning before failure.

This behaviour defines brittleness.

Brittleness does not mean weakness. Ceramics can support extremely high compressive loads and remain stable at temperatures where many metals soften or degrade. What they cannot tolerate is large tensile strain. When their elastic limit is reached, fracture occurs.

To understand brittleness, we must first examine how ceramics respond to stress.

When a load is applied to a ceramic body, the material initially responds elastically. Elastic behaviour means that the material deforms slightly under load and returns to its original shape when the load is removed. In ceramics this deformation is extremely small because the material is stiff. The structure resists strongly but stretches very little.

Figure 3.1 illustrates a fundamental difference in material behaviour.

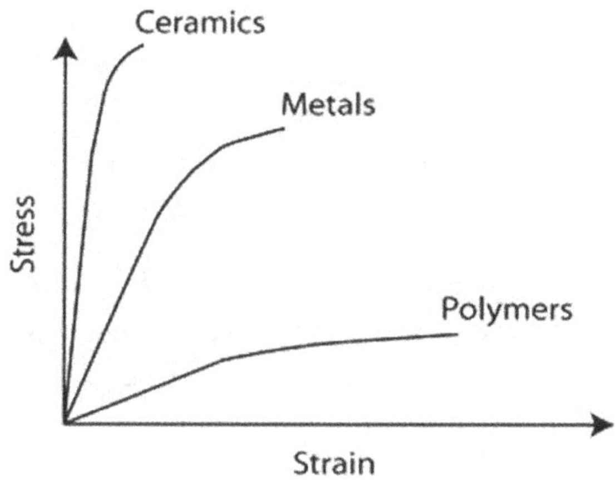

Figure 3.1 — Stress–Strain Behaviour: Ceramics vs Metals vs Polymers
Ceramics fracture after elastic deformation, while metals and polymers undergo plastic deformation before failure.

Ceramics exhibit high stiffness and brittle fracture with minimal plastic deformation. Metals display an elastic region followed by plastic deformation before fracture. Polymers have lower stiffness but can undergo very large strain before failure.

Metals typically deform in two stages. Under increasing load, they first respond elastically, then enter a plastic region where permanent deformation occurs. Plastic deformation redistributes stress, absorbs energy, and often provides visible warning before fracture.

Ceramics do not show this plastic stage. Their stress–strain response remains almost entirely linear until a critical limit is reached. When that limit is exceeded, atomic bonds break and a crack propagates rapidly through the material. Because plastic deformation does not occur to absorb energy or redistribute stress, fracture appears sudden.

This behaviour originates from the atomic bonding within ceramics. Ceramic materials are held together primarily by strong ionic and covalent bonds. These bonds are rigid and directional, preventing atomic layers from sliding past each other as they do in metals. When stress becomes too high, the bonds break rather than shift.

The same bond structure that produces high stiffness also produces brittleness.

Because stress cannot be relieved through plastic flow, ceramics become highly sensitive to defects. Small pores or microcracks can act as starting points for fracture.

When tensile stress is applied, these defects concentrate stress locally. Once the critical stress level is reached at the flaw tip, a crack begins to grow.

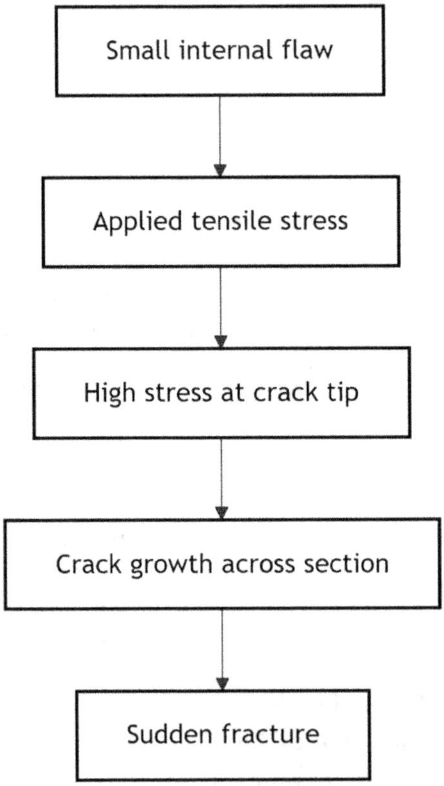

Figure 3.2 — Crack Initiation Under Tensile Stress
Tensile stress concentrates at flaw tips, initiating fracture in brittle ceramics.

Because ceramics lack plastic stress redistribution, crack growth can proceed very rapidly once initiated.

This mechanism explains many failures observed in ceramic components. Consider a ceramic tile subjected to bending. One surface of the tile experiences compression while the opposite surface experiences tension. If a flaw exists near the tensile surface, stress concentrates at that location. Once a crack begins, it can propagate rapidly across the section, producing a clean fracture surface because plastic deformation does not occur.

Bending tests clearly demonstrate this behaviour. In three-point bending experiments, fracture almost always begins on the tensile side of the specimen. Even when compressive stress is high on the opposite side, failure is controlled by tension. For this reason, bending strength tests are widely used for ceramics—they reveal both the tensile limit of the material and its sensitivity to flaws.

Brittleness also affects impact behaviour. Metals can absorb impact energy through deformation, but ceramics cannot. When an impact produces tensile stress beyond the critical level, fracture occurs.

This does not make ceramics unsuitable for structural applications. Instead, it means their use must be carefully controlled. Many ceramic systems are designed so that the material operates primarily in compression. Surface compression in tiles improves durability, and refractory linings are arranged to carry compressive loads while minimizing tensile stress.

Brittleness also influences geometric design. Sharp corners, sudden thickness changes, holes, and notches create stress concentration. Metals may relieve these stresses through plastic deformation, but ceramics cannot. Ceramic components must therefore be designed with smooth transitions and controlled geometry.

Ceramics are also very stiff materials. Even small deformation can generate significant stress when movement is restrained. For example, a tile rigidly fixed to a substrate may develop internal stress during temperature change. If the resulting tensile stress exceeds the fracture limit, cracking occurs.

Brittleness therefore implies three key characteristics:

- Limited tensile strain
- Absence of plastic deformation before fracture
- Rapid crack propagation once the critical stress level is reached

Reliable ceramic performance therefore requires careful control of geometry, loading conditions, and internal structure.

Elastic and Plastic Behaviour Compared

Table 3.1 — Elastic vs Plastic Deformation

Feature	Elastic Behaviour	Plastic Behaviour
Shape After Load Removal	Returns to original shape	Permanent deformation
Energy Absorption	Low	High
Stress Redistribution	Very limited	Significant
Present in Ceramics	Yes (small range)	No (negligible)
Present in Metals	Yes	Yes

This comparison explains why fracture behaviour differs strongly between ceramics and metals. Ceramics operate almost entirely within the elastic region. Once the elastic limit is exceeded, fracture occurs without a plastic stage to absorb energy or reduce stress concentration.

Understanding brittleness prepares us for the next topic: flaws and stress concentration. Because ceramics cannot redistribute stress plastically, the presence and size of defects ultimately determine where fracture begins and how much stress the material can tolerate.

Decision Insight:

- Ceramics fail when tensile stress interacts with a flaw.
- Smooth geometry reduces stress concentration.
- Controlling defects is more important than maximizing strength values.

Chapter 4

Flaws, Stress Concentration and Strength

A small defect can control the failure of a large component.

A ceramic does not fracture where the material is strongest. Fracture begins where stress encounters a critical flaw. While brittleness explains the sudden nature of ceramic fracture, it does not explain why identical materials can fail at different loads. The controlling factor is the size and location of defects within the stressed region.

Ceramic strength is not controlled by average quality. It is controlled by the most critical flaw inside the stressed region. This flaw may be so small that it cannot be seen with the naked eye. Yet it can decide the life of the entire component.

To understand this, we must understand stress concentration.

When a smooth ceramic body is pulled in tension, the stress ideally spreads evenly across the section. But real materials are never perfect. They contain pores, microcracks, inclusions, weak grain boundaries, or small surface scratches. These imperfections disturb the stress flow.

At the tip of a crack, stress does not remain average. It becomes much higher than the applied stress. The sharper the crack, the higher the stress at its tip. Even if the overall load appears safe, the local stress at the crack tip may already be at the breaking limit.

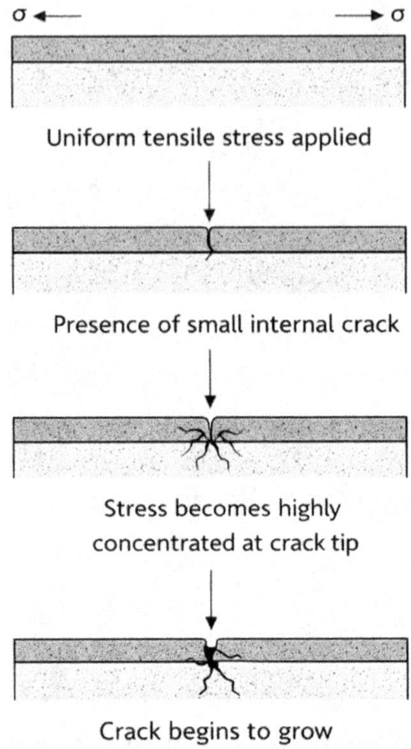

Figure 4.1 — Stress Concentration at a Crack Tip
Local stress amplification at the tip of a flaw under tensile loading.

Illustrates the basic concept of stress concentration at a crack tip. A crack tip behaves like a stress amplifier. In metals, plastic deformation near the crack tip can blunt the crack and reduce this amplification. In ceramics, however, plastic deformation is extremely limited. The crack tip remains sharp, and the stress concentration remains high.

When tensile stress is applied to a ceramic body, the stress is initially distributed relatively uniformly throughout the material. If a small internal crack or flaw is present, this distribution becomes highly uneven. The sharp crack tip concentrates the applied stress into a very small region. Once the local stress exceeds the fracture resistance of the material, crack growth begins.

This explains why ceramic strength is highly sensitive to flaw size. If the flaw is very small, the local stress may remain below the critical limit. If the flaw is slightly larger, the local stress can exceed the fracture limit even under moderate loading. Once crack propagation begins, it can spread rapidly across the section, leading to sudden fracture.

This leads to an important understanding: Strength decreases as flaw size increases.

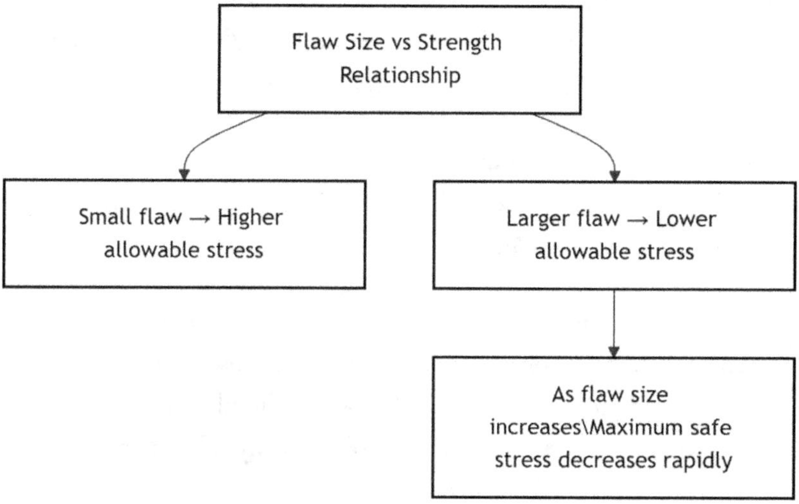

Figure 4.2 — Flaw Size vs Strength Relationship
Increasing flaw size increases stress concentration and reduces allowable strength.

The strength of ceramic materials is strongly controlled by the size of internal flaws.
Small flaws produce lower stress concentration and allow the material to withstand higher applied stress. Larger flaws create intense stress concentration at the crack tip, significantly reducing the safe stress level. As flaw size increases, the maximum allowable stress decreases rapidly, which explains the inherent strength variability in brittle ceramic materials.

Strength in ceramics is not a fixed number. It depends on the size of the largest defect under stress. A ceramic body therefore does not have one single strength value.

This explains the size effect. Small laboratory specimens often show higher strength than large industrial components. A small bar contains less volume and therefore a lower probability of a critical flaw. Large tiles or refractory blocks have greater volume and a higher chance of containing a larger defect somewhere inside.

In industrial production, many stages can introduce flaws. Mixing may trap air and form pores. Pressing can create density variation or lamination cracks. Drying may generate microcracks from uneven shrinkage. Firing and cooling can introduce internal stresses and cracks. Handling may create scratches or edge chips. Any of these defects can become a fracture origin.

Surface condition is especially important because tensile stress during bending is highest at the surface. Scratches behave like small cracks, so rough or damaged surfaces reduce bending strength.

For example, a large tile placed on uneven support experiences bending. The bottom surface enters tension. If a pore or edge chip exists near that tensile region, fracture may start there and spread quickly across the tile. In refractories, anchors can restrain expansion and create tensile stress locally. If a microcrack is present near the anchor region, spalling may begin.

Failure occurs when the stress concentrated at a flaw tip exceeds the fracture resistance of the material. The interaction between applied stress and flaw size therefore determines the effective strength of ceramic components.

This changes how strength should be viewed. The real question is not simply how strong the ceramic is, but what flaw size exists and what tensile stress it will experience. Reliability therefore depends on controlling defect size through good processing, careful handling, and thoughtful design.

Common Sources of Flaws in Ceramic Components

Table 4.1 — Typical Flaw Sources in Ceramic Components

Source Stage	**Typical Flaw Type**
Mixing	Trapped air pores
Forming	Lamination cracks, density variation
Drying	Shrinkage cracks
Firing	Uneven phase formation, thermal cracks
Cooling	Residual stress cracks
Handling	Surface scratches, edge chips
Installation	Local bending, restrained expansion

This table shows that flaws can enter at many stages. Reliability is not controlled at a single step. It is controlled throughout the process and the system.

Combining the concepts from the previous chapter clarifies the mechanism of ceramic fracture. Because ceramics cannot redistribute stress through plastic deformation, defects strongly amplify local stress. When the concentrated stress at a flaw exceeds the material's resistance to fracture, cracking begins.

Understanding flaws and stress concentration prepares us for deeper topics such as strength testing, reliability variation, and failure prediction.

Ceramic strength is not about how much load you apply.

It is about where that load concentrates — and what flaw is waiting there.

Because flaw formation is strongly influenced by processing and internal structure, understanding microstructure becomes essential for predicting ceramic reliability.

Decision Insight:

- If tensile stress cannot be eliminated, reduce flaw size.
- Surface flaws are critical where tensile stress occurs.
- Improved processing and surface quality increase reliability.

Chapter 5

Microstructure — The Hidden Architecture

Microstructure is the permanent record of processing decisions.

The shape you see is not the structure that decides performance. Two ceramic products can look identical from the outside. Same size. Same colour. Same finish. Yet one survives service and the other fails early. The difference is not visible on the surface. It lies inside.

That inside structure is called microstructure.

Microstructure is the internal architecture of a ceramic body. It includes grains, grain boundaries, pores, glassy phases, crystalline phases, and internal stress patterns. These features are small, but their influence is large. They control strength, expansion, thermal shock resistance, creep behaviour, and long-term durability.

Processing creates microstructure. Firing fixes it. Service tests it.

To understand ceramics, we must understand this hidden structure.

Ceramic bodies are not solid blocks of uniform material. They are made of many small crystals called grains. These grains grow and bond during firing. The regions where they meet are called grain boundaries. Between grains, there may be thin glassy phases formed from melted components during firing. There may also be pores left from incomplete densification or trapped air.

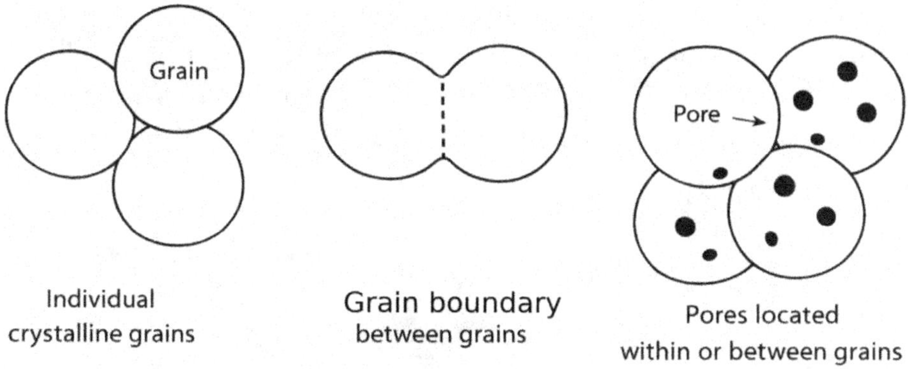

Figure 5.1 — Grain and Grain Boundary Structure
Ceramic bodies consist of crystalline grains separated by grain boundaries.

The basic internal pattern. Each grain is a small crystal. The boundaries are the joints between these crystals. The strength of these joints matters greatly.

Ceramic materials consist of many small crystalline grains formed during sintering. The interfaces where these grains meet are known as grain boundaries.
Small pores may exist either inside individual grains or along the grain boundaries. These microstructural features strongly influence mechanical strength, thermal behaviour, and fracture resistance of ceramics.

If grain boundaries are strong and clean, cracks find it harder to pass through. If boundaries contain glass pockets, impurities, or micro-voids, they become weak paths. Cracks prefer weak paths.

Grain size also changes behaviour. Fine grains create many boundaries. This can help deflect cracks and sometimes improve strength. Large grains reduce boundary area but may allow cracks to travel more directly. There is no single best grain size for all applications. It depends on service conditions.

For example, in refractories exposed to chemical attack, larger grains may resist corrosion better. In structural ceramics where strength is critical, finer and more uniform grains often improve reliability.

During firing, some components melt partially and form a glassy phase. This glassy phase flows between grains and helps bond them together. It fills gaps and reduces porosity. But too much glassy phase can reduce high-temperature strength. Glassy material softens more easily at elevated temperatures compared to crystalline phases.

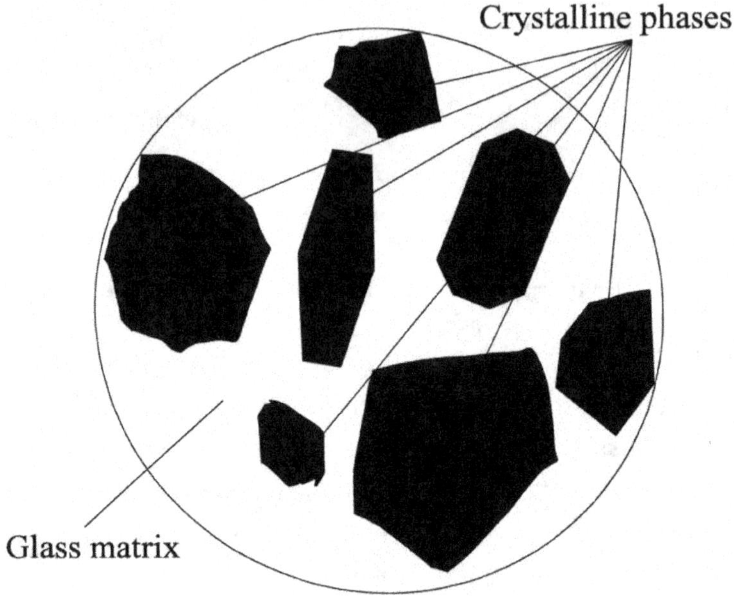

Figure 5.2 — Glassy vs Crystalline Phase Distribution
Glassy phase occupies grain boundaries and influences bonding and high-temperature behaviour.

Small pores may remain depending on densification and processing conditions.

Figure 5.2 illustrates this mixed structure, where the balance between crystalline and glassy phases is critical.

If the glassy phase is too low, bonding between grains becomes weak and strength decreases. If it is too high, creep resistance and high-temperature performance decline. In tiles, excess glassy phase can increase thermal expansion mismatch with the glaze. In refractories, it can reduce load-bearing ability at service temperature.

Porosity is another key element of microstructure. Pores reduce density and the effective load-bearing area. Under tensile stress, they act as stress concentrators, and larger or interconnected pores increase the likelihood of crack initiation. However, controlled porosity can be beneficial. In insulation refractories it reduces thermal conductivity, and in some cases a small amount of porosity can help absorb thermal stress. The important factor is control.

Microstructure also includes internal or residual stress. During cooling, different phases contract at different rates. If cooling is fast or uneven, stresses remain locked inside the body. These stresses may not be visible but reduce the margin between safe stress and fracture stress. For example, internal tensile stress from cooling combined with additional service stress can lead to sudden failure.

Because microstructure depends strongly on processing, small manufacturing variations can produce large differences in behaviour. Uneven compaction during pressing creates density variation, and during firing these regions shrink differently, producing uneven grain growth and porosity distribution. Cracks often initiate in the weaker regions.

Particle size distribution before forming also affects final structure. Fine particles pack and densify differently from coarse ones, and poor control can lead to non-uniform microstructure.

Microstructure ultimately controls performance. In tiles it influences bending strength, water absorption, and dimensional stability. In sanitaryware it affects glaze fit and crack resistance. In refractories it governs load resistance, creep behaviour, and chemical durability.

When strength, thermal shock resistance, or durability improves, it is usually because the microstructure has been refined. Microstructure links raw materials to real-world performance. In the Ceramic Behaviour System framework, it forms the bridge between processing and final properties.

Microstructural Features and Their Effects

Table 5.1 — Microstructural Features and Their Property Influence

Microstructural Feature	Effect on Behaviour
Small grain size	Often improves strength and crack deflection
Large grain size	May improve chemical resistance but reduce strength
High porosity	Lowers strength, may improve insulation
Low porosity	Increases density and strength
High glassy phase	Improves bonding, may reduce high-temperature stability
Strong grain boundaries	Improves resistance to crack growth
Residual tensile stress	Reduces safety margin against fracture

This table shows that microstructure is not decorative. It is decisive.

If microstructure is uneven, properties are uneven. If microstructure is controlled and uniform, behaviour becomes predictable.

This chapter reinforces a key link from Chapter 1: processing defines microstructure, and microstructure defines properties.

When a ceramic fails, the fracture surface often reveals the cause. It may show pores, weak boundaries, or glass-rich regions. Cracks simply follow the weakest path available.

Microstructure forms the internal skeleton of the ceramic. If this structure is balanced and strong, the material performs reliably. If it is uneven or weak, cracks can develop and grow.

The next chapter will examine how temperature changes interact with microstructure to create internal stress.

One principle remains clear: The performance of ceramics is written inside them long before they are used.

One of the most important consequences of microstructure and phase distribution appears during cooling — the development of residual stress.

🗒 Decision Insight:

- Processing builds the internal architecture of ceramics.
- Microstructure determines real properties.
- Hidden residual stresses reduce the safety margin.

End of Part I — How Ceramics Think

You now understand that ceramics are not unpredictable.
They are flaw-controlled, stress-sensitive, and structure-defined materials.
Everything that follows will apply these laws.
If you remember one idea from this part, remember this:
Performance begins long before failure appears.

Part II — Thermal Intelligence

Understanding How Temperature Creates Stress

Ceramics operate in environments where temperature change is constant. Heating, cooling, phase transformation, and thermal gradients all influence internal stress.

This section explains how thermal expansion, residual stress, phase transformations, and thermal shock affect ceramic stability. By understanding how temperature interacts with structure, engineers can prevent many of the most common failures encountered in ceramic production and service.

Chapter 6

Residual Stress — The Stress You Cannot See

Temperature change becomes dangerous when movement is restrained.

Some ceramics fail not because of the load they carry in service, but because of the load already locked inside them before use. A tile may leave the kiln looking flat and sound. A refractory block may pass inspection without visible defects. Yet inside the material, tensile or compressive stress may already exist. This hidden internal stress is called residual stress.

Residual stress is dangerous because it is invisible. It does not always produce immediate cracking. Instead, it reduces the safety margin of the ceramic. When additional stress is later applied during handling, installation, heating, cooling, or service, fracture may occur sooner than expected.

To understand residual stress, we must return to temperature and movement. During firing, ceramic bodies are heated to high temperature and expand naturally. Expansion itself is usually harmless. Problems begin during cooling. As temperature falls, the ceramic contracts. If contraction occurs uniformly and freely, stress does not develop. But in real production, cooling is rarely perfectly uniform.

The surface usually cools faster than the interior. The outer layer begins to contract while the inside remains hotter and more expanded. Because the hotter interior resists the contraction of the cooler surface, tensile stress develops near the surface.

Figure 6.1 illustrates this sequence. Larger temperature differences between surface and interior produce higher internal stress.

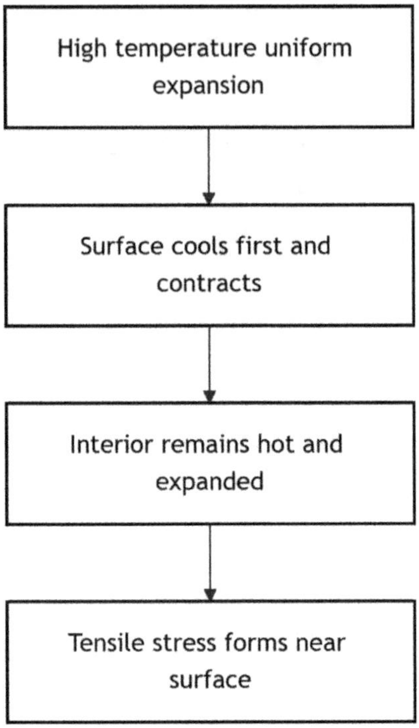

Figure 6.1 — Residual Stress Formation During Cooling
Faster surface contraction creates tensile stress when the interior remains hot.

Large tiles, thick sanitaryware sections, and heavy refractory blocks are especially sensitive because their interiors cool more slowly. If cooling is too fast, surface tension increases and cracks may form. Some cracks appear immediately. Others remain small and grow later.

Residual stress can also arise from internal phase differences. Most ceramics are not single-phase materials. They contain crystalline grains, glassy phases, and sometimes several crystal types. Each phase has its own thermal expansion behaviour. During cooling, one phase may attempt to contract more than another. Because the phases are bonded together, free movement is impossible, and stress develops at the interfaces. This phase mismatch may not cause immediate visible cracking, but it weakens the internal stress balance of the material.

In glazed ceramics, residual stress caused by glaze–body expansion mismatch is especially important. During cooling, both the ceramic body and the glaze contract. If the glaze contracts more than the body, the glaze is pulled into tension and surface crazing may occur. If the glaze contracts less than the body, the glaze is placed in compression. Controlled surface compression can improve crack resistance, but excessive compression may produce peeling, shivering, or edge lifting.

Figure 6.2 shows that mismatch alone can create stress even without any external load. The ceramic may appear intact after firing, but its internal stress condition has already been defined.

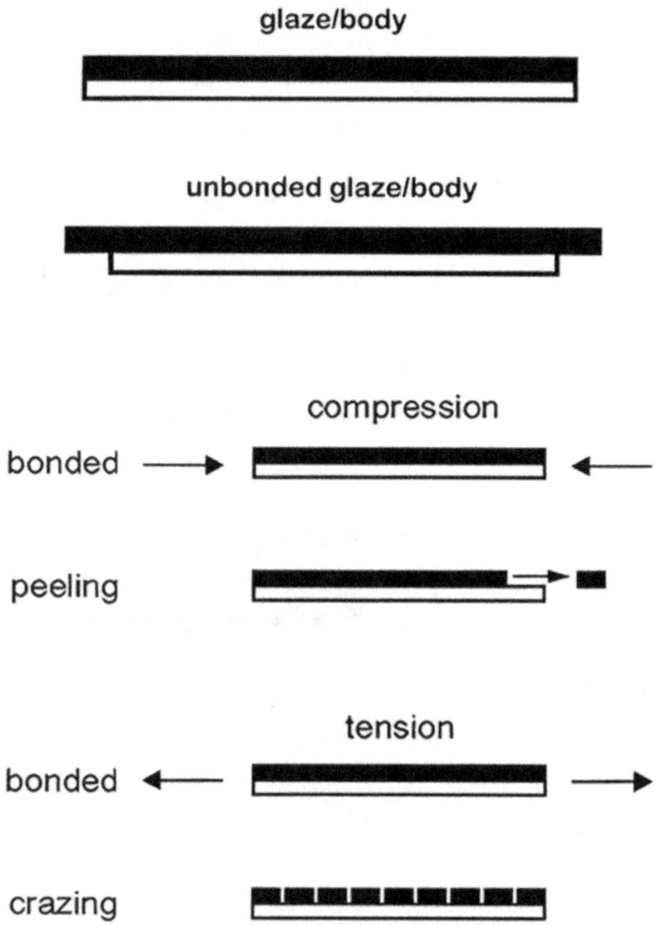

Figure 6.2 — Thermal Expansion Mismatch Between Body and Glaze

Glaze–body interaction showing compression (peeling risk) and tension (crazing formation) in glazed ceramics.

During cooling, the ceramic body and glaze contract at different rates. Because they are bonded together, this mismatch creates stress at the interface. Depending on the expansion relationship, the glaze may be placed in residual tension, leading to crazing, or in residual compression, which may improve durability or, if excessive, cause peeling or edge defects.

Residual stress also helps explain warpage. If one side of a tile cools faster than the other, uneven contraction causes curvature. Even when the product appears nearly flat after firing, internal stress may still remain. In refractories, cooling during shutdown is critical. Thick linings develop strong temperature gradients if cooling is too rapid. Surface layers contract while the interior remains hot, allowing cracks or microcracks to form. Even when visible spalling does not occur immediately, internal damage may shorten service life.

Residual stress is therefore closely linked to cooling rate. Slow, controlled cooling reduces thermal gradients and lowers stress. Proper kiln cooling curves influence not only maturity and phase development, but also the internal stress condition of the product. Component size also matters. Thin sections cool more uniformly than thick sections. Large sanitaryware items and refractory blocks require much stricter cooling control because their section thickness increases thermal imbalance.

Residual stress becomes even more dangerous when flaws are present. If a flaw already exists in a zone of tensile residual stress, the crack tip is stressed before service begins. Any additional service stress then brings the material closer to fracture.

Residual stress does not always cause immediate failure. More often, it reduces the difference between safe stress and fracture stress. Under repeated heating, cooling, or mechanical loading, microcracks may grow until visible damage appears.

Common Causes of Residual Stress

Table 6.1 — Common Causes of Residual Stress in Ceramics

Source	How Stress Forms
Rapid cooling	Surface contracts faster than interior
Thick sections	Interior remains hot longer
Phase transformation	Volume change during cooling
Glaze–body mismatch	Different thermal expansion rates
Uneven kiln temperature	Non-uniform contraction
Restrained shrinkage	Movement blocked during cooling

This table shows that residual stress is usually process-generated, not random. It develops through cooling gradients, phase change, mismatch, and restraint.

Residual stress reinforces an important principle from earlier chapters. Ceramics are sensitive to tension. They cannot relieve stress through plastic deformation. When tensile stress already exists inside the material, the safety margin is reduced before service even begins.

Residual stress is the stress you cannot see, but it must be controlled.

The next chapter will examine thermal expansion more directly. There we will see how temperature change creates movement, and how restrained movement becomes stress. But one truth must remain clear: what happens during cooling does not disappear. It remains inside the material and becomes part of its behaviour.

◈ Industrial Example 1 — Thermal Shock Cracks in Kiln Furniture

Situation:

Kiln setters developed cracks after rapid temperature change during kiln operation.

Initial assumption:
The refractory material contained a manufacturing defect.

Reality:
The material's thermal shock resistance was insufficient for the heating rate used in the kiln cycle.

Correction:
Select kiln furniture with higher thermal shock resistance and reduce the heating ramp to limit temperature gradients.

Lesson:
Heating rates must remain within the thermal tolerance of the material.

Explanation:
Thermal shock occurs when rapid temperature change creates steep temperature gradients inside the ceramic. The surface heats and expands faster than the interior, generating tensile stress. Cracks form once this stress exceeds fracture strength. Materials with lower thermal expansion, higher thermal conductivity, and greater fracture toughness tolerate larger gradients and show better thermal shock resistance.

Decision Insight:

- Rapid cooling creates hidden tensile stress.
- Expansion mismatch generates internal stress.
- Residual tension reduces fracture safety margin.

Chapter 7

Thermal Expansion — Movement That Creates Stress

Thermal movement becomes stress when expansion or contraction is restrained.

Ceramic materials are often exposed to large temperature changes during manufacturing and service. To understand cracking, distortion, and long-term failure, engineers must understand how ceramics move with temperature. Heat alone rarely breaks ceramics. Problems arise when thermal movement is uneven or restrained.

Every ceramic expands when heated and contracts when cooled. This movement follows basic physical law. If a ceramic body were completely free to move, expansion and contraction would not create stress. In practice, however, ceramic products are rarely free. They are bonded to glaze, fixed within structures, supported by frames, attached to metal parts, or restrained by adjacent components. When thermal movement is blocked or uneven, internal stress develops.

Thermal expansion is therefore one of the most important behaviours in ceramic systems. Within the Ceramic Behaviour System (CBS) framework, it forms a direct link between material properties and service conditions. As temperature rises, atomic vibration increases and the average atomic spacing becomes larger, causing expansion. As temperature falls, vibration decreases and contraction occurs. The amount of movement varies between materials, and even small differences can generate significant stress when temperature changes are large.

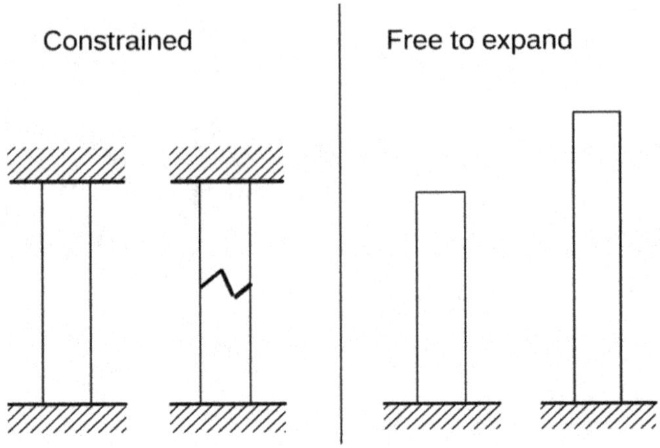

Figure 7.1 — Free Expansion vs Restrained Expansion
Restrained expansion generates internal stress during temperature change.

Illustrates the central principle of this chapter. Movement is natural. Restrained movement is dangerous.

Consider a large tile installed tightly between walls without an expansion joint. As temperature rises, the tile attempts to expand. If movement is blocked, compressive stress builds. If this stress becomes excessive, tenting or lifting may occur. During cooling, the reverse happens. The tile contracts. If contraction is restrained, tensile stress develops. If tensile stress exceeds the fracture limit, cracking begins.

Thermal expansion mismatch is also critical in layered systems. In glazed ceramics, the glaze and body are bonded together. During cooling after firing, both contract. If their thermal expansion values differ, internal stress develops at the interface.

If the glaze contracts more than the body, the glaze is pulled into tension and crazing may occur. If the glaze contracts less than the body, the glaze is placed into compression. Controlled compression can improve surface crack resistance, but excessive compression may cause peeling or edge-related defects.

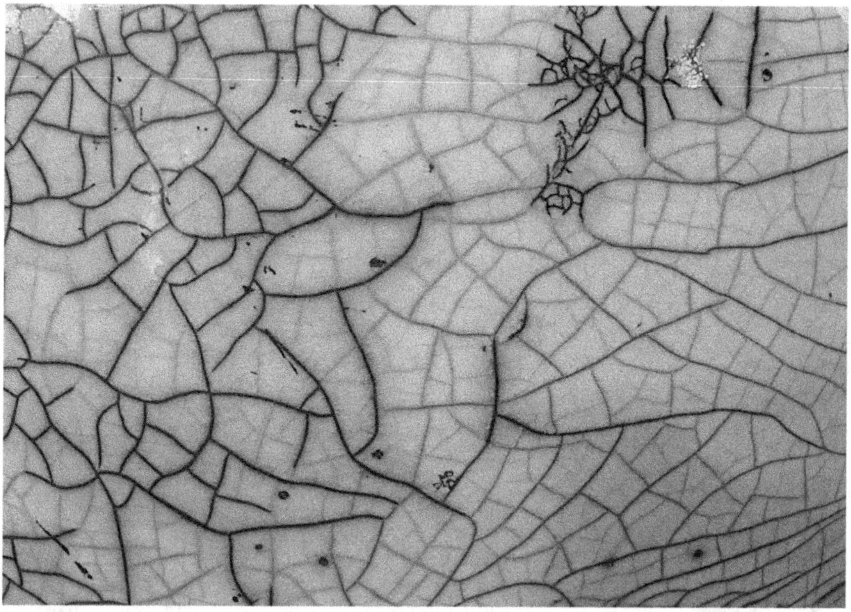

Figure 7.2 — Expansion Mismatch Between Two Bonded Materials
Bonded materials with different expansion coefficients generate interface stress.

Figure shows how two bonded layers with different expansion behaviour generate stress even without external mechanical load.

Thermal expansion is equally important in refractories. Furnace linings expand during heating. If sufficient expansion allowance is not designed into the system, compressive stress builds. During shutdown, cooling contraction may create tensile stress, especially where anchors, shells, or local constraints restrict movement.

Expansion behaviour must therefore be considered during design, not after installation. The magnitude of thermal movement depends on material type and temperature range. Some ceramics expand gradually. Others show abrupt dimensional change at certain temperatures because of phase transformations.

Large temperature changes produce larger movement. Rapid temperature change intensifies stress because temperature gradients form across the section thickness.

In thin tiles, temperature equalizes quickly. In thick refractory linings, one side may change temperature much faster than the other. This produces differential expansion through the thickness, and differential expansion becomes stress. Repeated heating and cooling further increase the problem, because cyclic movement can slowly extend microcracks over time.

Thermal expansion cannot be eliminated. It can only be managed. Design must allow movement. Installation must provide joints and clearance. Material selection must consider expansion compatibility wherever bonding or restraint exists. Ignoring thermal expansion does not stop movement — it simply converts movement into stress.

Typical Thermal Expansion Considerations

Table 7.1 — Thermal Expansion and Its Engineering Impact

Situation	**Risk Created**	**Control Method**
Large temperature rise	Compressive stress if restrained	Provide expansion allowance
Cooling from high temperature	Tensile stress if restrained	Controlled cooling rate
Glaze–body mismatch	Crazing or peeling	Match CTE values
Bonded dissimilar materials	Interface cracking	Compatible expansion design
Rapid temperature change	Thermal gradient stress	Gradual heating and cooling

This table shows that thermal expansion is not a theoretical issue. It directly affects glaze fit, installation, refractory layout, and service reliability.

As explained in Chapter 6, cooling contraction can generate residual stress inside ceramic bodies. Thermal expansion determines how much movement occurs during heating and cooling, while restraint determines how much stress develops.

Ceramics therefore do not fail because of temperature alone.
They fail when thermal movement is restrained and stress exceeds the fracture limit.

The next chapter examines thermal shock, where rapid temperature change creates severe internal gradients and sudden fracture risk. The principle remains simple:

Temperature change creates movement.
Restrained movement creates stress.
Stress controls fracture.

◈ Industrial Example 2 — Tile Corner Cracking After Firing

Situation:
Corner cracks appeared in tiles immediately after firing.

Initial assumption:
The cracks were caused by rough handling.

Reality:
Rapid cooling created tensile stress at the corners, where stress concentration is naturally higher.

Correction:
Control the cooling rate through the critical temperature range.

Lesson:
Cooling is an engineered stage of firing and must be controlled carefully.

Explanation:
During cooling, the outer surfaces contract before the interior fully cools. This creates tensile stress, especially at corners and edges where geometry intensifies local stress. Because ceramics are weak in tension, these regions become likely crack initiation sites. Controlled cooling allows gradients to relax more gradually and reduces corner cracking.

⟐ Industrial Example 3 — Crazing in Large-Format Ceramic Tiles

Situation:
Fine crack networks (crazing) appeared on the glaze surface of large-format ceramic tiles after firing or during service.

Initial assumption:
The glaze layer contained manufacturing defects or surface damage.

Reality:
Thermal expansion mismatch between the glaze and ceramic body placed the glaze in tensile stress during cooling.

Correction:
Adjust glaze composition and match the coefficient of thermal expansion (CTE) of the glaze to the body to maintain slight surface compression.

Lesson:
Glaze–body expansion compatibility must be controlled to prevent tensile stress in the glaze.

Explanation:
During cooling after firing, both the glaze and ceramic body contract. If the glaze contracts more than the body, tensile stress develops in the glaze layer. Because ceramics are sensitive to tensile stress, this tension can produce fine crack networks known as crazing. Matching glaze expansion to body expansion allows the glaze to remain in slight compression, improving surface durability and preventing crack formation.

Decision Insight:

- Thermal expansion always creates movement.
- Restrained movement converts movement into stress.
- Expansion allowance must be designed, not assumed.

Chapter 8

Thermal Shock — When Temperature Changes Too Fast

Thermal shock occurs when rapid temperature change creates internal stress faster than the material can tolerate.

Many ceramics are designed to operate at high temperature. Refractories hold molten metal. Kiln furniture survives repeated firing. Tiles tolerate normal temperature variation. Heat alone is not usually the problem. The real danger appears when temperature changes too quickly.

Rapid heating or cooling produces temperature gradients inside the ceramic body. One region changes temperature before another. The surface may expand or contract while the interior lags behind. This difference in movement creates internal stress. If the resulting tensile stress exceeds the material's fracture strength, cracking occurs. This is thermal shock.

Thermal shock is therefore not about the absolute temperature. It is about how uneven the temperature becomes inside the body. Temperature gradients create internal stress. Speed increases the difference. The faster the temperature change, the greater the risk.

Consider rapid cooling first. When a hot ceramic is exposed suddenly to cold air or water, the surface cools and contracts immediately. The interior remains hotter and more expanded. The hot interior resists the contraction of the cooler surface, and tensile stress develops at the surface.

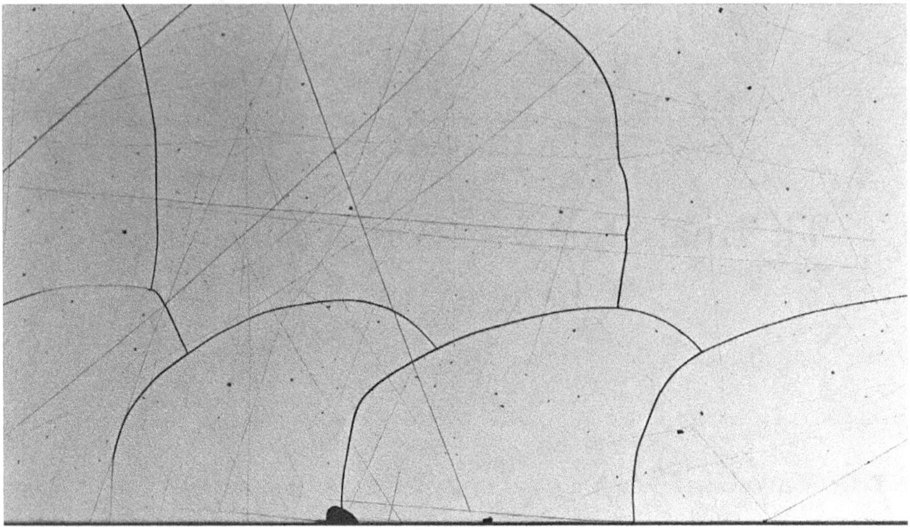

Figure 8.1 — Rapid Cooling and Surface Tensile Stress
Surface tensile stress generated by rapid cooling gradients

Figure explains why sudden cooling often produces surface cracking. The surface becomes the weak zone because tension develops there.

Rapid heating can also be dangerous. If a cold ceramic is exposed suddenly to high temperature, the surface heats and expands before the interior does. The resulting mismatch creates internal stress. In thick refractory linings, this is common during furnace start-up. The hot face expands while the cold face lags behind, producing internal bending stress. If heating is too aggressive, cracks can form through the lining.

Thermal shock resistance depends on several material properties acting together.

Table 8.1 — Factors Controlling Thermal Shock Resistance

Property	Influence on Thermal Shock Resistance	Engineering Meaning
Thermal Expansion (CTE)	Lower expansion reduces stress during temperature change	Materials with low CTE tolerate larger temperature gradients
Thermal Conductivity	Higher conductivity spreads heat faster	Reduces temperature difference between surface and interior
Tensile Strength	Higher strength delays crack initiation	Material tolerates higher stress before fracture
Fracture Toughness	Higher toughness slows crack propagation	Small cracks are less likely to grow catastrophically
Elastic Modulus	Lower modulus reduces stress for same strain	More compliant materials tolerate thermal strain better
Microstructure	Controlled porosity can improve thermal shock resistance by reducing crack driving force	Microstructural design influences crack growth behaviour

These properties work together. Low thermal expansion reduces dimensional change. High thermal conductivity reduces thermal gradient. High tensile strength delays crack initiation. Greater fracture toughness slows crack propagation. Microstructure determines how easily cracks begin and spread.

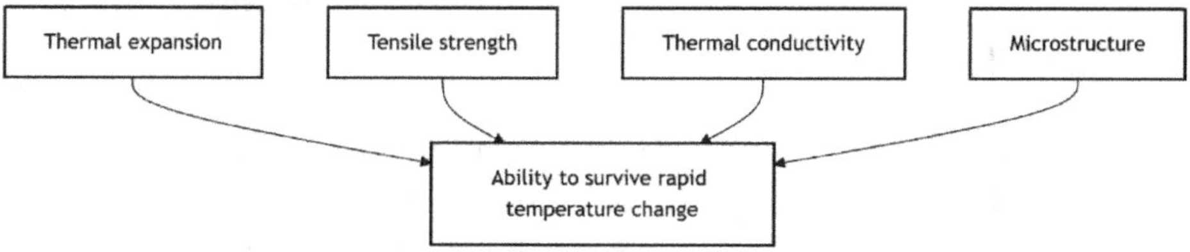

Figure 8.2 — Factors Affecting Thermal Shock Resistance

Thermal shock resistance is controlled by the combined effects of thermal expansion, tensile strength, thermal conductivity, and microstructure.

Good thermal conductivity reduces temperature gradients and therefore lowers thermal stress. Microstructure (grain size, porosity, and crack-deflecting features) further influences how easily cracks initiate and propagate. Together, these factors determine a ceramic's ability to survive sudden temperature change.

Thermal shock resistance is controlled by multiple material properties acting together. Low thermal expansion reduces strain during rapid heating or cooling, while higher tensile strength increases crack resistance.

Figure shows that thermal shock resistance is never controlled by one property alone. A dense ceramic with high strength but high thermal expansion may still fail under rapid cooling. A refractory with moderate strength but lower expansion and balanced porosity may survive repeated thermal gradients more successfully.

Geometry also matters. Thick sections are more vulnerable because the temperature difference between surface and interior becomes larger. Thin sections equalize temperature faster and therefore develop lower thermal stress. Size matters as well. A small laboratory specimen may survive a severe test, while a large industrial block of the same material may crack because its internal thermal gradient is greater.

Thermal shock damage often begins at the surface. Small cracks may form first and remain unnoticed. Repeated exposure can extend them further until pieces begin to detach. In refractories, this leads to spalling.

Thermal shock appears in many industrial situations. In tiles, it may develop when hot water contacts a cooler surface. In sanitaryware, uneven heating may create local stress concentration. In refractories, rapid shutdown or aggressive start-up often creates crack networks near the hot face.

Control begins with temperature management. Gradual heating and cooling reduce gradients. Preheating reduces sudden mismatch. Expansion joints and proper support conditions reduce restraint. Material selection is equally important. Materials with lower thermal expansion, good conductivity, and balanced microstructure usually show better thermal shock resistance.

Thermal shock is closely linked with brittleness, flaws, and residual stress. Because ceramics lack plastic stress redistribution, rapid temperature change can quickly push local stress beyond the fracture limit.

Ceramics do not fail because they are exposed to high temperature. They fail because temperature changes faster than stress can be safely managed.

The next chapter will move from single rapid events to repeated thermal loading. There we will examine how thermal cycling slowly accumulates damage over time. But one principle is already clear: temperature difference creates stress. Temperature speed multiplies it. And stress controls fracture.

◈ Industrial Example 4 — Cooling Cracks in Sanitaryware

Situation:
Cracks developed in sanitaryware products during the cooling stage near the exit of tunnel kilns.

Initial assumption:
The defects were caused by material weakness or forming defects in the ceramic body.

Reality:
Rapid cooling near the kiln exit created strong temperature gradients between the surface and the interior of the product.

Correction:
Adjust the kiln cooling profile and introduce controlled cooling zones to reduce temperature gradients.

Lesson:
Cooling rate must be controlled in thick ceramic sections to prevent thermal stress during contraction.

Explanation:

During rapid cooling, the outer surface of the ceramic contracts before the interior temperature has decreased. The hotter interior resists this contraction, generating tensile stress near the surface. Because ceramics are weak in tension, this stress can exceed the fracture limit and produce cracking. Controlled cooling allows temperature differences to equalize gradually, reducing thermal stress and preventing fracture.

Decision Insight:

- Rapid temperature change creates internal gradients.
- Temperature gradients generate tensile stress.
- Thermal shock occurs when stress exceeds strength.

Chapter 9

Thermal Cycling and Fatigue — Damage That Grows Slowly

Repeated temperature change gradually accumulates damage inside ceramic materials.

Ceramics do not always fail from one severe event. Often, they fail from repeated moderate events. A component may survive one heating and cooling cycle without visible damage. It may survive ten cycles or even hundreds. But after enough repetition, cracks begin to appear. This is thermal fatigue.

Thermal cycling means repeated temperature change. Every cycle produces expansion during heating and contraction during cooling. If that movement is restrained or uneven, stress develops in every cycle. Even when the stress in a single cycle is too low to cause immediate fracture, it can still create small damage. Over time, this damage accumulates.

Ceramics are brittle. They cannot absorb much stress by deformation. Small flaws can therefore grow slowly under repeated loading, even when each individual cycle remains below the immediate fracture limit. This slow crack growth is one of the major causes of delayed ceramic failure.

During heating, certain regions enter stress because of expansion mismatch or restraint. During cooling, contraction produces stress again. The stress state is not always identical in both directions. One region may be most critical during heating, another during cooling. Repeated stress reversal gradually weakens the structure.

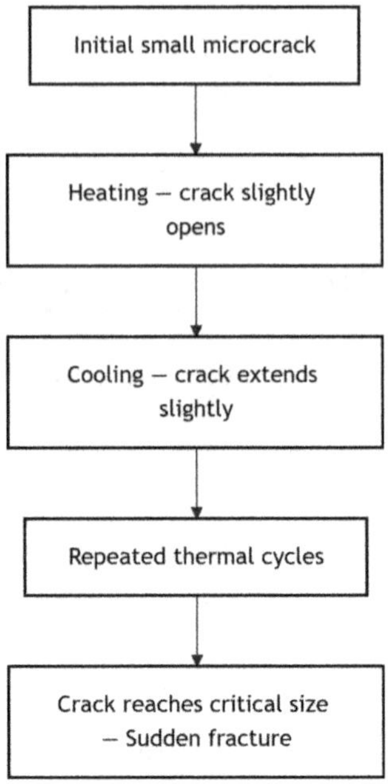

Figure 9.1 — Crack Growth Under Repeated Thermal Cycles
Repeated heating and cooling gradually extend microcracks.

Figure shows how crack growth under cycling is incremental. The crack does not usually cross the section in one event. It extends step by step. Each cycle may add only a very small amount of damage, but after enough cycles the crack reaches critical size.

Once that critical size is reached, one additional cycle may cause full fracture. This explains why components can fail after months or years of service even though they passed initial testing. In some ceramic systems, stress corrosion and environmental effects can also assist this slow crack growth at stresses below the immediate fracture limit.

Thermal fatigue is common in refractories used in batch-operated furnaces. Every start-up and shutdown cycle produces expansion and contraction. Because the surface changes temperature faster than the interior, thermal gradients develop in each cycle. Microcracks form at the hot face and slowly grow deeper. Eventually spalling occurs.

The same behaviour appears in other ceramic products. Kiln shelves may appear sound after many firings, yet edge cracks grow slowly until sudden fracture occurs under normal load. In tiles, repeated temperature changes from sunlight, hot water, or service exposure can produce gradual crack development, especially if installation restrains movement.

Several factors control the severity of thermal fatigue. Larger temperature swings create larger movement and higher stress. Rapid heating or cooling increases thermal gradients. High thermal expansion increases stress per cycle. Low tensile strength reduces the tolerance for damage. Residual tensile stress from manufacturing reduces the available safety margin. Geometry also matters. Corners, edges, holes, thickness transitions, and anchor zones act as stress concentrators where cracks often begin.

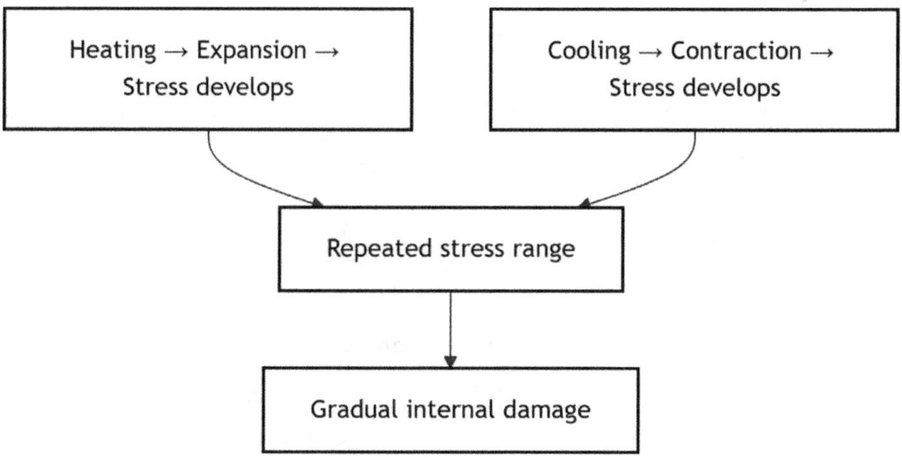

Figure 9.2 — Stress Range During Thermal Cycling
Repeated heating and cooling create alternating tensile and compressive stresses that gradually extend microcracks.

Figure makes an important point: fatigue depends not only on peak stress, but on repeated stress range. Even when peak stress stays below the immediate fracture strength, repeated cycling can still grow cracks.

This is why single-cycle strength testing cannot fully predict service life. A ceramic may show high laboratory bending strength and still fail in service if it experiences repeated thermal cycling.

Thermal fatigue damage is often progressive and initially invisible. Surface may show fine crack lines. In refractories, small flakes may detach. In advanced stages, larger sections may spall away. The danger is that failure appears sudden, but damage has been growing slowly for a long time.

Improving resistance to thermal fatigue requires reducing the stress range and slowing crack growth. Lower thermal expansion reduces movement per cycle. Controlled heating and cooling reduce thermal gradients. Expansion joints allow freer movement. Improved microstructure slows crack propagation. Gradual start-up and shutdown procedures are especially important in industrial systems.

Thermal fatigue connects directly to earlier chapters. Brittleness means there is no plastic stress relief. Flaws amplify local stress. Thermal expansion creates movement. Thermal shock creates severe gradients. Residual stress reduces safety margin. Repetition then grows damage until fracture occurs.

A ceramic does not need one extreme event to fail. Repeated moderate stress can be equally destructive.

These thermal mechanisms ultimately act through one controlling parameter: the tensile strength of the ceramic.

Understanding what strength actually means in brittle materials is therefore essential, which we examine in the next chapter.

Factors Influencing Thermal Fatigue

Table 9.1 — Thermal Cycling and Fatigue Risk Factors

Factor	Effect on Fatigue Life
Large temperature swing	Shortens life
Rapid heating or cooling	Increases crack growth rate
High thermal expansion	Increases stress per cycle
Low tensile strength	Reduces tolerance
Residual tensile stress	Accelerates failure
Stress concentration zones	Crack initiation sites
Frequent cycling	Accumulates damage faster

This table shows that thermal fatigue is predictable. It depends on stress range, material behaviour, flaw condition, and repetition.

Ceramics may survive one cycle without damage. They may survive many cycles. But repeated stress leaves a structural memory. That memory grows into cracks. And eventually, fracture follows.

◈ Industrial Example 5 — Edge Chipping During Handling

Situation:
Frequent edge damage occurred during packing and transportation of tiles.

Initial assumption:
The damage was caused only by rough handling.

Reality:
Tile edges had lower mechanical strength because of insufficient densification at the edges during forming and firing.

Correction:

Improve pressing uniformity and optimize firing to achieve more consistent densification across the tile.

Lesson:

Mechanical integrity begins in the forming stage.

Explanation:

During pressing, density is rarely perfectly uniform. Edges and corners often receive lower compaction, leading to higher porosity after firing. These weaker regions chip easily when tiles contact each other or hard surfaces. Better density uniformity improves edge strength and reduces transport damage.

Decision Insight:

- Thermal cycling damage accumulates.
- Repeated stress grows hidden cracks.
- Fatigue resistance must be designed, not assumed.

End of Part II — Thermal Intelligence

Temperature is not dangerous by itself. Uneven temperature under restraint is. Thermal expansion, residual stress, thermal shock, and thermal cycling are all connected through one central rule: movement becomes dangerous when it is restricted or uneven.

Engineering thermal behaviour is therefore engineering survival.

Part III — Mechanical Truth

Understanding Strength, Fracture, and Time Effects

Mechanical behaviour in ceramics differs significantly from that of ductile materials. Strength is governed by flaws, not average material properties, and failure can occur without visible warning.

This part examines how strength is measured, why size and probability influence failure, and how time-dependent processes such as creep and slow crack growth affect reliability. It also explains the limitations of laboratory testing when applied to real industrial systems.

Chapter 10

Strength in Ceramics — What It Really Means

Ceramic strength is statistical — governed by flaw size.

The strength number on a report is not a promise. It is a result under specific conditions. When a ceramic product is evaluated, one of the first numbers people look at is strength. A test report may show a bending strength value and that number often becomes the basis for approval.

But in ceramics, strength is not as simple as it looks. Strength is not a fixed property like density. It is not a guaranteed capacity. It is the stress level at which a particular sample fractured under a particular test condition.

To understand ceramic strength properly, we must understand how it is measured and what controls it. Most ceramic strength values come from bending tests. In a typical three-point bending test, a ceramic bar is supported at two ends and loaded at the centre.

The top surface experiences compression. The bottom surface experiences tension. Because ceramics are weak in tension, fracture almost always begins at the tensile surface.

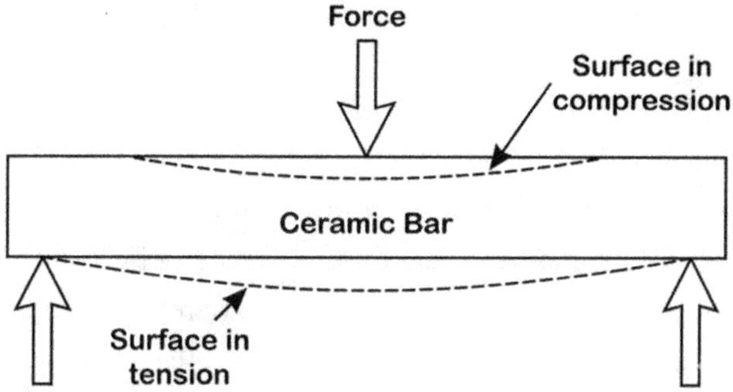

Figure 10.1 — Bending Test and Tensile Failure Zone
In bending, fracture usually begins on the surface under tensile stress.

Figure shows why bending is used. It exposes the tensile limit of the material, which is usually the controlling factor in failure.

The stress at fracture in bending is called the Modulus of Rupture, or MOR. It is widely used in ceramic industries. However, MOR is not a fundamental constant. It depends on flaw size, specimen size, surface quality, and test setup.

If two samples are made from the same batch, they may show slightly different MOR values because the largest flaw inside each sample may be different. This variation is normal. Ceramic strength is statistical because flaws are statistical.

This leads to an important idea: strength decreases with increasing volume.

A small laboratory specimen has a lower chance of containing a large critical flaw. A large tile or refractory block has a higher chance of containing one. Even if the material is the same, the probability of a large defect increases with size. This is why large industrial components may fail at lower stress than small test bars.

Another important factor is temperature. Strength measured at room temperature is called cold MOR. Strength measured at operating temperature is called hot MOR.

In many ceramics, hot strength is lower than cold strength. At elevated temperature, glassy phases may soften. Grain boundaries may weaken. Creep may begin. If a product operates at high temperature, cold MOR alone is not enough for evaluation.

Ceramic strength measured at room temperature is often higher because the material structure remains stable and crack growth is limited. At elevated temperatures, strength may decrease due to thermal effects such as microcrack activation, grain boundary weakening, or creep. Therefore, test conditions should represent the actual service temperature to obtain reliable strength data.

A material may show high cold strength but may not perform well at service temperature.

Loading mode also affects strength interpretation. Compressive strength of ceramics is usually much higher than tensile or bending strength. But most real failures involve tension created by bending, restraint, or thermal gradients.

If a ceramic component is mainly under compression, it may perform very well. But if even small tensile zones appear due to design or installation, fracture risk increases.

Surface condition is another critical factor. Because bending stress is highest at the surface, surface defects are highly dangerous. A smooth, polished test sample may show high strength. A real product with small scratches, edge chips, or handling damage may show lower effective strength.

Machining marks in advanced ceramics can reduce strength significantly if not properly controlled.

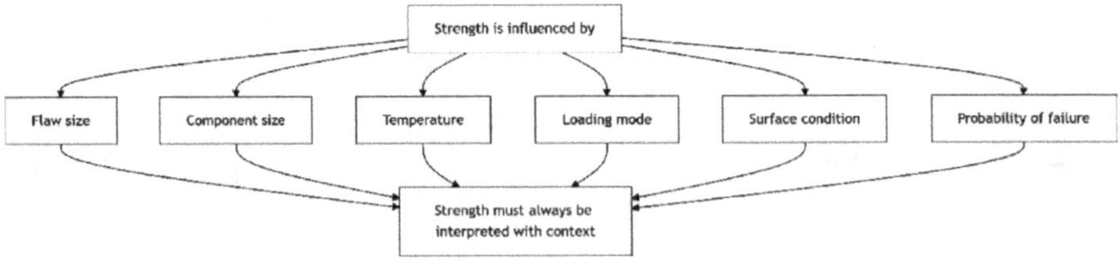

Figure 10.2 — Factors Influencing Strength

Strength depends on flaw size, stress distribution, material structure, and test geometry.

When reviewing a test report, asking only "What is the MOR value?" is not enough.

We must also ask:

At what temperature was it measured?
What was the sample size?
How many samples were tested?
What is the service stress condition?
Is tension present in the real application?

Without this context, strength numbers can be misleading. Strength in ceramics is not an absolute guarantee. It is the stress at which a particular flawed sample fractured under specific conditions.

Interpreting Ceramic Strength

Table 10.1 — Strength Interpretation in Ceramics

Strength Concept	Practical Meaning
Modulus of Rupture (MOR)	Stress at fracture in bending test
Cold MOR	Strength measured at room temperature
Hot MOR	Strength measured at operating temperature
Compressive strength	Much higher than tensile strength
Statistical variation	Strength differs between samples
Surface condition effect	Scratches and defects reduce strength

This table summarizes what strength really means.

In ceramics, we do not design based on average strength alone. We design with safety margin because flaw distribution cannot be perfectly controlled.

A high strength value is useful. But understanding how that value was obtained is more important. Ceramics do not fail because the reported strength was low.

They fail when tensile stress at a flaw exceeds the material's real tolerance under service conditions.

Ceramics are strong in compression but weak in tension. Even small tensile stresses can cause crack initiation and rapid failure. This is why ceramic components must be designed to minimize tensile loading.

The next chapter will examine the limitations of testing and how laboratory results must be interpreted carefully before applying them in real systems.

But one rule must remain clear. Strength in ceramics is conditional. It depends on flaws, temperature, stress type, and probability. And it must always be interpreted in context.

🏭 Industrial Example 6 — Strength Variation in Ceramic Tile Testing

Situation:
Ceramic tiles from the same production batch showed different breaking strengths during quality testing.

Initial assumption:
The variation was caused by inconsistent processing conditions or material defects in production.

Reality:
Even when processing conditions are identical, microscopic flaws inside the ceramic structure vary from piece to piece.

Correction:
Interpret strength results statistically and maintain strict control over processing steps that influence defect formation.

Lesson:
Ceramic strength is controlled by the largest flaw present in the stressed region, not by the average material quality.

Explanation:

Ceramic materials are flaw-controlled. During strength testing, fracture occurs at the location of the most critical defect within the stressed area. Because defect size and location vary slightly between individual pieces, measured strength values also vary. For this reason, ceramic strength is typically described using statistical distributions rather than a single fixed value. Understanding this behaviour helps engineers interpret test data and reliability more realistically.

Decision Insight:

- MOR is a test result, not a guaranteed strength.
- Larger products fail earlier because flaw probability increases.
- Surface damage can reduce real strength dramatically.
- Design must minimize tensile stress zones.

Chapter 11

Testing Limitations — When Numbers Mislead

Strength in ceramics is a statistical outcome, not a fixed property.

A test result describes a sample. It does not guarantee the system. Testing is necessary in ceramic engineering. Without testing, we cannot compare materials or control production quality. But testing has limits. The mistake is not in testing itself. The mistake happens when numbers are accepted without understanding their conditions.

In ceramics, this misunderstanding is common. A laboratory report may show a high bending strength value. That number looks strong and reassuring. But the real question is not "What is the number?" The real question is "Under what conditions was this number obtained?"

Most ceramic strength tests are performed under controlled laboratory conditions. Specimens are small. Surfaces are smooth. Load is applied slowly and evenly. Temperature is constant. Support conditions are ideal. Real life is rarely like that.

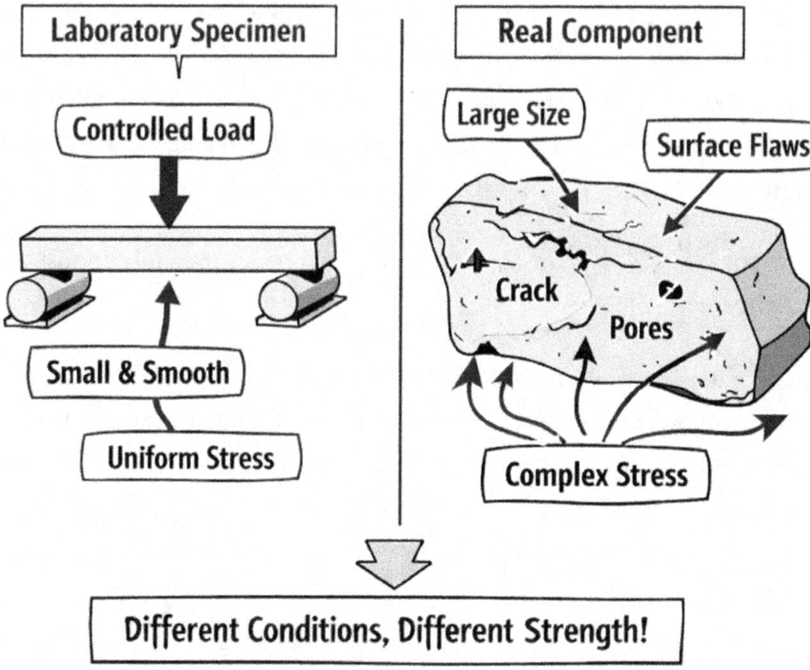

Figure 11.1 — Ideal Laboratory Test vs Real Component
Laboratory specimens often experience more uniform stress than real ceramic components.

Figure shows the gap between laboratory testing and real service.

A small laboratory bar has a lower probability of containing a large flaw. A large industrial tile or refractory block has a higher probability. The chemistry may be identical, but the risk level is different.

Laboratory bending tests create a very specific stress pattern. Maximum stress exists only at one surface location. In real applications, stress may come from bending, thermal expansion, restraint, impact, or combined loading. Stress may not be evenly distributed.

Loading rate is another important difference. In laboratory tests, load is usually applied slowly. In service, load may be sudden. Thermal shock, impact during handling, or installation stress can act much faster. Fast loading leaves less time for any internal stress redistribution, even small effects.

Temperature condition is also critical. Many strength values are measured at room temperature. But if the product operates at high temperature, hot strength is more relevant. Glassy phases soften at elevated temperatures. Grain boundaries may weaken. A ceramic that is strong at room temperature may behave differently at 800°C or 1200°C.

If only cold MOR is considered in high-temperature applications, risk is underestimated.

Statistical variation is another limitation. Ceramic strength is not a single fixed number. It follows a distribution. Each specimen has its own largest flaw. Some samples fail earlier. Some fail later.

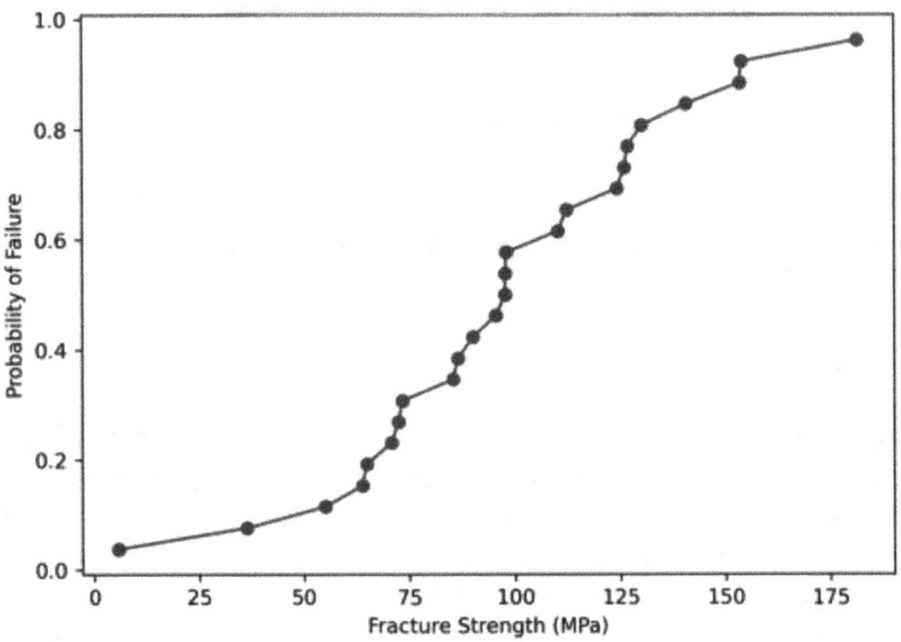

Figure 11.2 — Strength Distribution in Ceramic Samples
Because strength is flaw-controlled, ceramic samples show statistical strength variation.

Figure shows that strength values spread over a range. The average value hides weaker samples.

If design uses only average strength without safety margin, lower-end pieces may fail in service.

Surface condition also changes strength. Laboratory samples are often carefully polished or cut. Real products may have minor edge chips from cutting, transport damage, or installation scratches. These surface defects act as crack starters, especially in bending.

Testing rarely includes long-term cycling unless specifically designed for it. A product may pass a static strength test but fail after repeated thermal cycling. Single-load tests do not always represent fatigue behaviour.

Testing compressive strength is another example of misunderstanding. Ceramic compressive strength is often very high. But most real failures occur in tension. Relying only on compressive values gives false confidence.

Testing is a tool. It is not the final answer. The purpose of testing is to provide controlled comparison and quality control. It helps detect production variation. It helps compare materials under identical conditions. But it cannot replace understanding of service environment.

When reviewing test data, we must ask deeper questions.

Does the test simulate the real stress type?
Does the temperature match operating conditions?
Is component size comparable?
Was surface condition realistic?
Is statistical variation considered?

If the answer to these questions is no, interpretation must adjust accordingly. Engineering judgment connects laboratory data to real performance.

Common Testing Misinterpretations

Table 11.1 — Testing Limitations in Ceramic Evaluation

Test Result	Possible Misinterpretation	Correct Approach
High cold MOR	Product is strong in all conditions	Verify hot strength and service temperature
High compressive strength	Safe under all loads	Evaluate tensile stress zones
Small specimen testing	Represents large components	Consider size effect and flaw probability
Smooth laboratory surface	Real surface strength is equal	Account for handling and edge damage
Single-load test passed	Long-term reliability ensured	Evaluate thermal cycling and fatigue
Average value used	All pieces perform equally	Apply safety margin for variation

This table reminds us that numbers require interpretation.

Testing is essential. Without it, quality control is impossible. But numbers must be understood in context. Ceramic engineering does not end at the test machine. It begins when we connect test results to real stress, real temperature, real geometry, and real time.

A test result is not the material. It is one measured response under controlled conditions.

If we forget that, numbers mislead. If we remember it, testing becomes a powerful decision tool.

Decision Insight:

- Cold strength does not represent service strength.
- MOR is a comparison value, not a guarantee.
- Ceramic strength follows statistical variation.

Chapter 12

Size Effect — Why Bigger Is Not Stronger

Larger components carry greater risk because flaws scale with size.

In brittle materials, size increases probability — not capacity. In many materials, increasing size increases strength because more material means more load-bearing area. In ceramics, the opposite often happens.

A small ceramic test bar may show high strength. A large tile made from the same batch may fail at lower stress. This is not inconsistency. It is a fundamental rule of brittle fracture.

Ceramics are flaw-controlled. They do not fail because the average structure is weak. They fail because a critical flaw exists in a stressed region.

When the component is small, the volume under stress is small. The probability of containing a large flaw is lower. When the component is large, the stressed volume increases. With increased volume comes increased probability that a larger defect exists somewhere inside that region.

The material has not changed. The probability has.

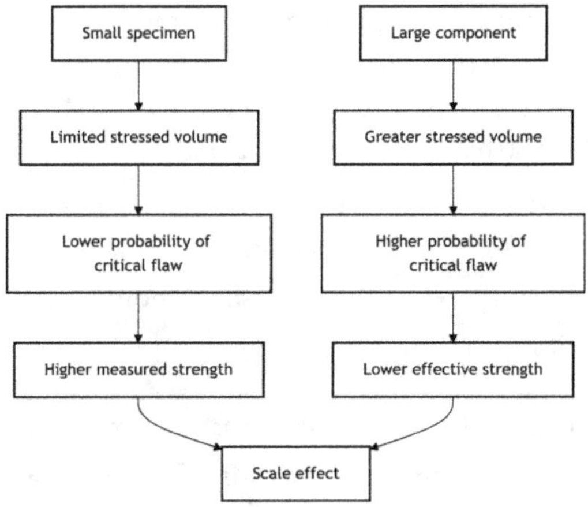

Figure 12.1 — Scale Effect Illustration (Small vs Large Component)
Larger components have higher probability of containing critical flaws and therefore show lower average strength.

A larger component has a greater stressed volume, increasing the likelihood of a critical flaw being present. As a result, the effective strength of large ceramic components is typically lower than that measured on small test specimens.

Figure shows the logic clearly.

Imagine a ceramic body containing randomly distributed micro-defects. In a small test specimen, none of these defects may be large enough to cause early failure. In a larger tile or block, however, the probability that at least one larger flaw exists becomes higher. Failure therefore begins at the largest critical flaw within the stressed zone. For this reason, strength in ceramics is statistical.

This explains several common industrial observations. Large-format tiles often show lower effective bending strength than small laboratory bars. Thick refractory blocks may crack even though sample testing indicated sufficient strength. Large kiln furniture components may fail earlier than expected. The reason is not necessarily poor production quality; it is the increase in stressed volume.

Surface area also contributes to the size effect. Surface flaws are especially critical because tensile stress is often highest at the surface. As surface area increases, the probability of scratches, edge chips, or machining marks also increases. Handling damage therefore becomes more significant in large components, with edges and corners acting as stress concentrators.

Geometry variations also become more influential in large bodies. Small differences in thickness, local density, or microstructure can affect stress distribution more strongly as size increases. Thermal gradients in large refractories or uneven support in large tiles can create wider tensile regions, increasing the likelihood that a critical flaw lies within the stressed area.

Because of this statistical behaviour, safety factors must increase when moving from laboratory samples to industrial components. Testing a few small specimens cannot fully represent the behaviour of thousands of installed pieces. Even a low defect probability becomes significant in large installations.

Unlike ductile materials, ceramics cannot redistribute stress through plastic deformation. Stress remains concentrated and fracture follows once a critical flaw becomes active. Increasing size therefore increases exposure to flaw probability.

Large ceramic components can still be reliable, but design, quality control, and safety margins must account for statistical behaviour. Reducing defect size, improving uniformity, and careful processing and handling all help reduce the severity of size effects, though the probability principle always remains.

Engineering Meaning of Size Effect

Table 12.1 — Size Effect and Design Implications

Factor	**Small Specimen**	**Large Component**
Stressed volume	Limited	Large
Probability of critical flaw	Lower	Higher
Measured strength	Higher	Lower
Strength variability	Smaller range	Larger range
Required safety margin	Moderate	Increased
Sensitivity to surface defects	Lower	Higher

This table shows the practical consequences.

If laboratory MOR is 50 MPa, it does not guarantee a large tile will safely sustain 50 MPa in service. Real allowable stress must account for size effect.

Everything connects back to the ten laws of ceramic behaviour.

Ceramics are flaw-controlled.
They do not yield before fracture.
Stress concentrates at defects.

When size increases, defect probability increases. Size does not change the chemistry. It changes the statistics. And in brittle materials, statistics govern failure.

The Ceramic Decision Framework

Engineering decisions involving ceramics often require balancing multiple material properties simultaneously. A ceramic that performs well in one condition may fail under another if the controlling property changes.

Selecting or evaluating ceramic materials therefore requires identifying the dominant stress condition.

The following framework summarises the most common service scenarios and the material characteristics that control performance.

Service Condition	**Dominant Risk**	**Critical Property**
Mechanical load	Fracture	Tensile strength, flaw population
Rapid temperature change	Thermal shock	Expansion coefficient, strength, conductivity
Sustained high temperature	Creep	Grain structure, glassy phase
Chemical exposure	Corrosion	Phase stability
Thermal cycling	Fatigue	Crack resistance

Each condition emphasises a different aspect of CBS.

Under mechanical loading, the controlling factor is flaw size. Ceramics fail when tensile stress interacts with critical defects. Strength therefore depends on flaw distribution rather than average composition.

Under rapid temperature change, thermal gradients create stress within the body. Thermal shock resistance depends on the combination of thermal expansion, elastic modulus, and fracture strength.

At high temperature, creep becomes significant. Materials containing glassy phases or fine grain boundaries may deform slowly under load.

Chemical environments introduce another dimension. Slag attack, oxidation, or corrosion can weaken structure even when mechanical stresses remain moderate.

Finally, thermal cycling subjects materials to repeated stress fluctuations. Microcracks can grow gradually until failure occurs.

The practical implication is simple:

No single property defines ceramic performance. Engineers must evaluate ceramics through the lens of the dominant service condition. Correct identification of the controlling mechanism often prevents incorrect material selection.

This framework provides a quick reference for aligning material behaviour with operating conditions.

◈ Industrial Example 7 — Warpage in Large Format Tiles

Situation:
Tiles developed centre lift after firing.

Initial Assumption:
Uneven kiln setters were suspected to be causing distortion.

Reality:
Differential shrinkage occurred between the tile surface and interior during sintering.

Correction:

Optimize particle size distribution and maintain a uniform firing curve to ensure balanced densification throughout the tile body.

Lesson:

Densification gradients control geometry.

Explanation:

During sintering, ceramic particles bond and the body shrinks as porosity is reduced. If densification occurs unevenly between the surface and the interior, different regions of the tile shrink by different amounts. This imbalance generates internal stress and causes geometric distortion such as centre lift or edge curl.

Large format tiles are particularly sensitive to these effects because small differences in shrinkage become amplified over larger dimensions. Controlling particle packing, density uniformity, and firing temperature distribution helps maintain balanced shrinkage and prevents warpage.

Decision Insight:

- Larger components have higher failure probability.
- Uniform flaw distribution improves reliability.
- Ceramic reliability is statistical, not absolute.

Chapter 13

Time Dependence — When Strength Changes with Duration

Stress acting over time slowly reduces structural safety.

In ceramics, time is not passive. It slowly reshapes strength. When we test a ceramic sample in a laboratory, the load is applied for a short period. The sample fractures, and the test ends. The number obtained represents strength at that moment.

But real components do not experience load for seconds. They experience load for months, years, sometimes decades. In ceramics, time changes behaviour.

A component that survives a short test may fail under the same stress if that stress continues long enough. This is not because the material suddenly becomes weak. It is because internal processes continue under stress.

Two major time-dependent mechanisms control this behaviour:

1. Creep under sustained load at high temperature
2. Subcritical crack growth under sustained tensile stress

Both are slow. Both are progressive. Both are often invisible until failure occurs.

Creep — Slow Deformation Under Heat and Load

Creep is slow deformation that occurs when a material is exposed to sustained load at elevated temperature.

At room temperature, most ceramics show almost no creep. Their atomic structure is rigid. But at high temperature, especially when glassy phases are present, atomic movement becomes easier. Grain boundaries may slide slowly. Viscous flow may begin in glassy regions.

Under constant load, strain increases gradually with time.

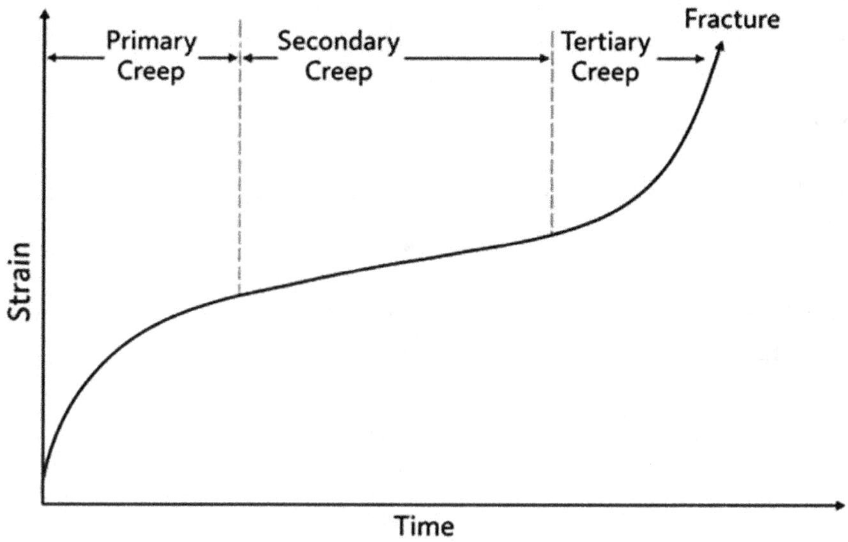

Figure 13.1 — Creep Curve
Time-dependent deformation occurs under sustained load at elevated temperature.

Figure shows the typical creep curve.

In Stage 1, deformation begins and slows slightly as the structure adjusts.
In Stage 2, deformation continues at a steady rate.
In Stage 3, damage accumulates rapidly and failure occurs.

In refractories, creep is critical. A lining under sustained mechanical load at high temperature may slowly deform. Bricks may compress. Joints may open. Structural alignment may change. Even without sudden fracture, dimensional stability can be lost.

In kiln furniture, long exposure to high firing temperatures under load may cause gradual sagging. The material may not crack immediately. It may simply lose shape.

Creep rate increases with:

Higher temperature
Higher applied load
Higher glassy phase content
Lower density

Dense, well-sintered microstructures resist creep better than porous ones under load.

This means cold MOR alone is not sufficient for high-temperature applications. Hot strength and long-term creep resistance must be considered.

A refractory may survive initial heating. But after thousands of operating hours, creep may compromise its structural role. Time slowly converts load into deformation.

Subcritical Crack Growth — Slow Crack Extension

Not all time-dependent failure involves deformation. Sometimes, no visible deformation occurs before fracture.

Subcritical crack growth happens when a crack grows slowly under stress lower than the immediate fracture strength. A small crack may exist inside the ceramic. The applied stress may be below the value required for instant fracture. But if that stress remains for a long time, the crack tip experiences concentrated stress continuously. Over time, atomic bonds at the crack tip slowly break.

The crack extends slightly.

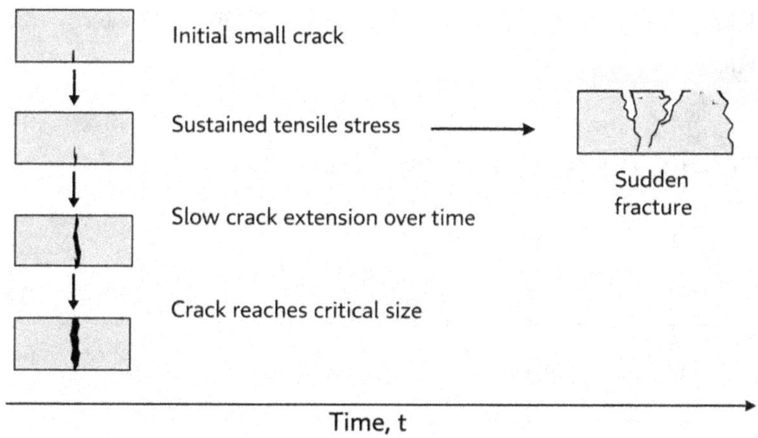

Figure 13.2 — Subcritical Crack Growth Concept
Cracks can slowly extend under stress levels below immediate fracture strength.

Figure shows the concept clearly.

The applied stress does not change. The crack size changes. Eventually, the crack becomes large enough that the same stress causes instant fracture. This explains delayed failure.

A ceramic component may appear stable for months. Then one day, under normal operating load, it fractures. The load did not increase. The crack grew.

Moisture and environment can accelerate subcritical crack growth. In humid conditions, certain grain boundaries become more vulnerable. At elevated temperature, atomic mobility increases, accelerating crack extension.

This mechanism is particularly important in tiles under sustained bending stress, refractory anchors under tensile load, and ceramic components exposed to constant pressure. Time reduces strength indirectly by increasing flaw size.

Time and Thermal Cycling

Time also interacts with thermal cycling. Repeated heating and cooling may not cause immediate failure. But each cycle slightly extends microcracks. Over long duration, cumulative damage leads to fracture.

Thermal fatigue is therefore a time-dependent process. Short-term tests rarely capture this effect.

Engineering Meaning of Time Dependence

Table 13.1 — Time Dependent Behaviour in Ceramics

Mechanism	Condition	Long-Term Effect
Creep	High temperature + sustained load	Gradual deformation
Subcritical crack growth	Sustained tensile stress	Delayed fracture
Thermal fatigue	Repeated temperature cycles	Progressive cracking
Environmental interaction	Moisture or chemical exposure	Accelerated crack growth
Long service duration	Constant moderate stress	Reduced safety margin

This table shows that time modifies behaviour even when stress does not change.

Design must consider:

How long will the component carry load?
At what temperature?
Under what environment?
How many thermal cycles?

A stress that is safe for one hour may not be safe for one year.

Time quietly reduces safety margin. Time connects directly to earlier the ten laws of ceramic behaviour.

Ceramics are flaw-controlled.
They do not yield before fracture.
Stress concentrates at crack tips.

If stress remains long enough, cracks grow. If temperature remains high enough, creep occurs. Time is not passive. It quietly reduces safety margin.

And when the margin becomes zero, fracture appears sudden — even though damage has been growing for a long time.

📦 Decision Insight:

- Sustained stress slowly grows cracks.
- High temperature enables creep deformation.
- Time reduces the safety margin against fracture.

Chapter 14

Process–Property–Performance — The Irreversible Chain

Laboratory strength values do not guarantee plant performance.

A ceramic does not fail in service. It completes a decision made during processing. When a ceramic component cracks in service, attention usually turns to the operating condition. Load may be blamed. Temperature may be blamed. Installation may be blamed.

But in ceramics, performance is rarely decided at the moment of use. It is decided much earlier. It is decided during processing.

Ceramics are not assembled like metal parts. They are created through a sequence of transformations. Powder becomes compact. Compact becomes green body. Green body becomes sintered structure. That structure becomes fixed.

Once the structure is formed, it cannot be reconfigured. The relationship between process, property, and performance is linear and irreversible.

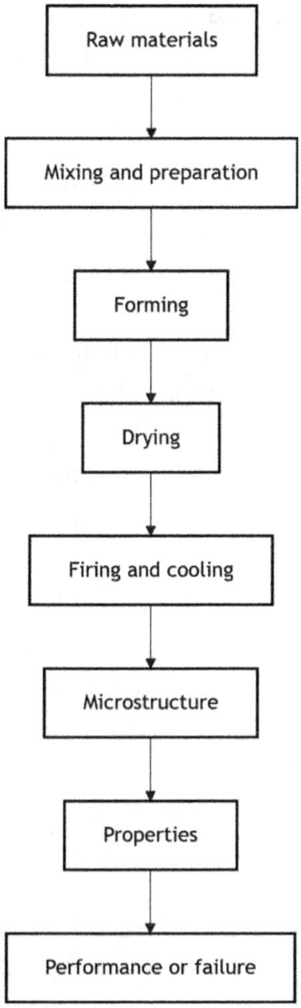

Figure 14.1 — Process-Property-Performance Chain
Processing conditions determine microstructure, which governs properties and service behaviour.

Figure shows the structural chain clearly.

Mixing, forming, drying, firing, and cooling determine the final microstructure of the material.

The resulting microstructure controls the material properties, which ultimately govern the performance of the ceramic component in service or its potential failure.

Raw materials define chemistry, particle size distribution, impurity content, and mineral structure. These factors influence packing behaviour and later phase formation.

Mixing controls homogeneity. Poor mixing creates local composition variations that may lead to expansion mismatch or weak zones.

Forming determines density distribution. Pressing pressure, slip rheology, or extrusion parameters control how uniformly particles are packed. Uneven density leads to uneven shrinkage during drying and firing.

Drying controls shrinkage gradients. If moisture leaves the surface too quickly, tensile stresses develop and microcracks may form inside the body.

Firing creates bonding. Sintering forms grain connections, phases transform, glassy phases develop, and pores shrink. Insufficient firing leaves weak bonding, while excessive firing may produce too much glass and reduce high-temperature strength.

Cooling determines residual stress. Uneven cooling or phase transformations can create stresses that remain locked inside the structure.

At the end of cooling, the microstructure becomes fixed. Properties such as strength, density, porosity, thermal expansion, creep resistance, and thermal shock resistance are established. Service performance is simply the response of this structure under operating conditions.

The process chain does not allow later correction. Lamination cracks from pressing do not heal in firing. Density variation from forming cannot be corrected later. Residual stresses created during cooling remain within the material.

Inspection can reveal visible defects but cannot remove internal weaknesses. Quality in ceramics is therefore built step by step, not created by final inspection.

Each process stage becomes a structural decision point. Raw material impurities may cause expansion mismatch, inconsistent forming pressure creates density variation, uneven drying produces microcracks, unstable firing alters phase development, and rapid cooling increases residual stress.

Ceramics record their process history in the microstructure.

Failure analysis therefore works backward. When a component cracks, investigation begins with the microstructure and traces through firing, drying, forming, and raw materials.

Performance is the final expression of all earlier decisions.

Engineering Meaning of the Process Chain

Table 14.1 — Process Influence on Final Performance

Process Stage	Structural Effect	Possible Failure if Uncontrolled
Raw material selection	Chemistry and particle characteristics	Expansion mismatch, weak bonding
Mixing	Homogeneity	Local weak zones
Forming	Density distribution	Lamination cracks, uneven shrinkage
Drying	Moisture removal gradient	Microcracks, warpage
Firing	Phase formation and bonding	Low strength, excess glassy phase
Cooling	Residual stress formation	Delayed cracking
Installation	Constraint and support	Tensile stress concentration

This table reinforces the chain.

Failure often appears in service, but its origin lies upstream.

Process–property–performance logic connects directly with earlier the ten laws of ceramic behaviour.

Ceramics are flaw-controlled.
Flaws originate during processing.
Microstructure defines properties.
Properties respond to stress and temperature.

The chain is continuous and irreversible.

If the process is stable, performance becomes predictable.
If the process is inconsistent, variability increases.
If variability increases, probability of failure increases.

Ceramics do not change after firing. They reveal what was built into them. Performance is not an accident. It is the outcome of processing discipline.

The next chapter will move into forming behaviour and green strength — where structural decisions first begin to solidify. But one principle must remain clear.

In ceramics, process is not preparation. Process is destiny.

Designing with Ceramics — Engineering for Brittle Materials

Ceramics perform extremely well when their behaviour is respected. Most failures do not occur because ceramics are weak; they occur because the design assumes the material behaves like metal.

Design errors often begin with the wrong material assumption.

Metals bend before they break, while ceramics do not. Metals redistribute stress through plastic deformation, whereas ceramics concentrate stress at flaws. Because of this difference, design philosophy must change.

Designing with ceramics requires controlling tensile stress, avoiding stress concentration, and allowing movement during temperature change. Ceramic components should never be designed with the expectation that the material will absorb structural mistakes. Ceramics reward careful geometry and controlled loading.

The First Principle: Reduce Tensile Stress

Ceramic failure is governed primarily by tensile stress. Compressive stress tends to close cracks, while tensile stress causes cracks to open and grow.

For this reason, good ceramic design minimizes tensile stress wherever possible. This principle appears in many applications:

- Tiles perform best when supported uniformly from below.
- Refractory linings carry load mainly in compression.
- Ceramic rollers are designed to minimize bending stress.
- Furnace structures allow expansion to prevent tensile restraint.

When tension cannot be avoided, defect control becomes critical.

The Second Principle: Avoid Stress Concentration

Stress concentration is a frequent cause of ceramic failure. Sharp geometry creates localized stress amplification where even moderate loads produce very high stresses.

Typical stress concentration features include:

- Sharp corners
- Notches
- Sudden thickness changes
- Holes and slots
- Edge defects

Smooth transitions distribute stress more evenly. Often a small geometric improvement increases reliability more effectively than increasing tensile fracture strength.

The Third Principle: Control Thickness Transitions

Ceramics respond poorly to sudden structural changes. Abrupt thickness changes produce uneven stress distribution and larger thermal gradients.

Sudden section changes may produce:

- Thermal stress during heating and cooling
- Bending stress under load
- Shrinkage differences during firing

Gradual transitions help maintain uniform stress flow and reduce local stress peaks.

The Fourth Principle: Provide Expansion Allowance

Ceramics expand when heated and contract when cooled. If expansion is restrained, internal stress develops and cracking may occur.

Typical situations include:

- Tiles installed without expansion joints
- Refractory linings constrained by steel shells
- Anchors restricting movement
- Large panels fixed rigidly on all sides

Expansion joints, flexible supports, and proper anchor design allow natural movement. Movement itself is not dangerous—restrained movement is.

The Fifth Principle: Design the Support System

Ceramics operate within mechanical systems. Supports, anchors, frames, and installation conditions influence stress distribution.

Poor support conditions can introduce bending stress even in strong materials. Examples include:

- Uneven tile bedding causing localized bending
- Incorrect anchor spacing in refractory linings
- Point loading instead of distributed support
- Large unsupported spans

Uniform support reduces tensile zones and improves durability.

The Sixth Principle: Respect Size Effects

Ceramic strength decreases as component size increases because larger parts have a higher probability of containing critical defects. This phenomenon is known as the size effect.

A small laboratory specimen may show high strength, while a larger industrial component made from the same material may fail at lower stress.

Design must therefore consider:

- component size
- defect probability
- stress distribution within the volume

Safety margins must reflect real component geometry rather than laboratory samples.

Design Discipline for Ceramic Components

Design Rule	Engineering Reason
Minimize tensile stress	Tensile stress opens cracks
Avoid sharp corners	Reduces stress concentration
Use gradual thickness transitions	Prevents local stress peaks
Provide expansion allowance	Prevents thermal stress
Ensure uniform support	Reduces bending stress
Consider size effect	Larger components contain larger flaws

Design is Part of Reliability

Ceramic reliability depends not only on material quality but also on design discipline. A well-designed component with moderate material strength can perform reliably for years, while a poorly designed one may fail early even if the material quality is high.

The difference lies in how the material is used. Ceramics do not tolerate design shortcuts; they respond directly to stress, structure, and constraint. When these factors are respected, ceramics become highly reliable engineering materials.

Decision Insight:

- Processing decisions determine final structure.
- Microstructure fixes material properties.
- Performance is the outcome of the process chain.

End of Part III — Mechanical Truth

Strength is not a single number.
It is statistical, size-dependent, and time-sensitive.
Testing informs design, but it does not replace understanding.
Reliability improves when flaw control improves.

Part IV — Density & Porosity

Understanding Internal Structure and Its Consequences

Density and porosity determine many of the physical and mechanical properties of ceramic materials. Small variations in internal structure can significantly influence strength, permeability, and thermal behaviour.

This section explains how density is measured, how pore structures develop during processing, and why uniform compaction and firing are critical for consistent performance.

Chapter 15

Density Defined — What "Solid" Really Means

Density measurements describe structure only when properly understood.

In ceramics, density is more than a physical number. It is a structural indicator. When we hold a ceramic product in the hand, we often judge it by weight. A heavier piece may appear stronger, but weight alone does not define density. Density is the relationship between mass and volume, and in ceramics that relationship reveals how much real solid structure exists inside the body.

A ceramic with high density contains more solid material within a given volume. A ceramic with low density contains more empty space in the form of pores. This difference influences strength, durability, thermal behaviour, and long-term reliability. Density therefore is not simply a measure of heaviness. It reflects how effectively the internal structure has been formed.

During ceramic processing, density records several structural events at once. It reflects how well particles were packed during forming, how effectively they bonded during firing, and how much pore space remains after sintering. For this reason, density is one of the simplest measurements with the deepest structural meaning.

To interpret density correctly, three related terms must be clearly distinguished: Bulk density, Apparent density, and True density.

Bulk density considers the total external volume of the specimen, including the pore space inside it. It therefore represents the compactness of the body as it actually exists.

Apparent density excludes open pores that can be filled with water during testing. This measurement reflects the density of the solid structure together with closed pores, giving a different perspective on internal compactness.

True density refers only to the density of the solid material itself, excluding all pores. It represents the theoretical density of the ceramic composition.

These terms are often confused, and this confusion leads to incorrect interpretation. In ceramics, a density value is meaningful only when the type of density being reported is clearly defined.

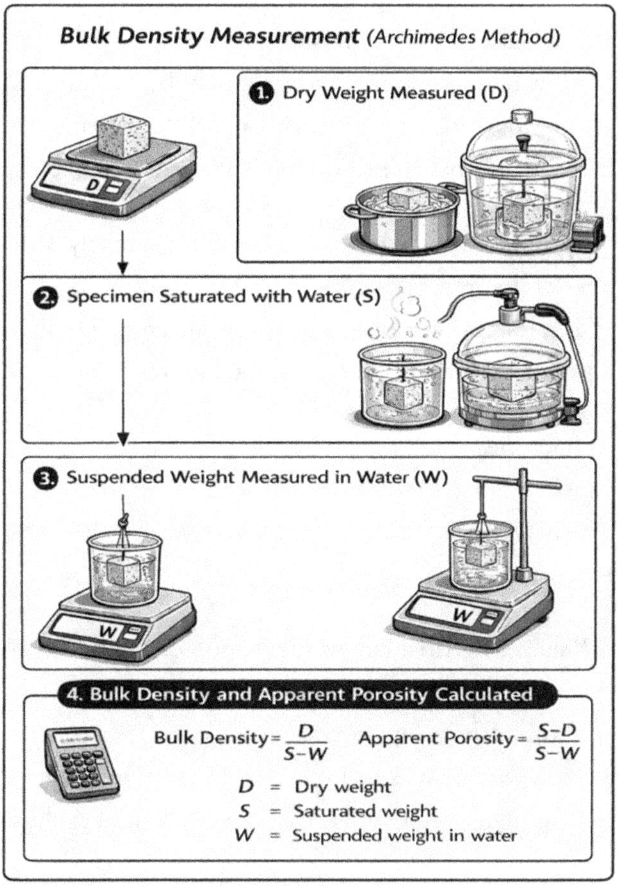

Figure 15.1 — Density Measurement Schematic

Illustration of the principle used to determine bulk density by relating mass to the measured volume of the ceramic body.

The measurement method itself is simple, but the information it provides is highly important. By comparing dry weight, saturated weight, and suspended weight, it becomes possible to estimate how much of the body is solid structure and how much is pore space.

In industrial practice, density testing is routine, yet its significance is often underestimated. If bulk density falls below specification, the cause may originate from several stages of processing. Poor particle packing during forming, insufficient compaction pressure, incomplete sintering during firing, or excessive residual porosity may all reduce density.

Variation in density also provides important warnings. If density varies widely between production batches, process control may be unstable. If density varies within a single component, uneven shrinkage during firing may occur. Such imbalance can create internal stress, warpage, or reduced reliability.

Density is closely linked to mechanical behaviour. In many structural ceramics and tiles, higher density generally improves strength because more solid material carries the load and fewer pores act as stress concentrators. However, density must always be interpreted within the context of the intended application.

In insulating refractories, for example, lower density is intentionally created to reduce heat transfer. Some strength is sacrificed to improve thermal insulation. A dense body is therefore not always better. The correct density depends on the service requirement.

Density also influences thermal behaviour. Dense ceramics usually conduct heat more efficiently, while porous ceramics reduce heat flow. At elevated temperatures, dense refractories often show improved resistance to creep compared with highly porous bodies. However, extremely high density combined with excessive glassy phase formation can reduce hot strength or create thermal expansion imbalance.

As with many ceramic properties, density must be understood as part of a broader structural balance.

Density gradients within a component are especially problematic. If the surface is denser than the core, or if different regions are compacted unevenly, shrinkage during firing becomes non-uniform. This imbalance may lead to warpage, residual stress, or cracking. Such gradients commonly arise from non-uniform pressing in tiles or uneven deposition in slip-cast bodies.

Density therefore acts as a fingerprint of both forming quality and firing effectiveness. With a relatively simple measurement, it reveals how much solid structure truly exists inside the ceramic.

Engineering Meaning of Density

Table 15.1 — Density and Performance Relationship

Density Condition	**Structural Meaning**	**Engineering Effect**
High bulk density	Low porosity	Higher strength
Low bulk density	High porosity	Lower strength
Density gradient	Non-uniform shrinkage	Internal stress risk
Under-fired body	Incomplete bonding	Weak structure
Over-fired body	Excess glassy phase	Reduced high-temperature stability
Controlled low density	Designed porosity	Improved insulation

This summary shows the practical meaning of density in ceramics.

Density is not related to strength alone. It also influences thermal conductivity, creep resistance, permeability, and dimensional stability. These effects connect directly to the Ten Laws of Ceramic Behaviour discussed earlier in the book.

Ceramics are flaw-controlled materials, and pores behave as flaws. Because ceramics do not yield before fracture, the amount and distribution of pore space strongly influence reliability. Density therefore reflects how many internal stress concentrators exist within the structure.

Density also records the quality of processing. Effective mixing improves particle packing. Proper forming increases structural uniformity. Correct firing strengthens bonding and reduces residual porosity. The measured density is therefore not an isolated number—it is the outcome of processing decisions.

In ceramics, density expresses how much solid structure exists inside the material, and structure ultimately governs performance and service life.

The next chapter moves from density to porosity. Density tells us how much solid structure exists, while porosity explains the architecture of the empty space that remains.

Decision Insight:

- Measure density correctly before interpreting it.
- Density indicates structure, not strength alone.
- Always specify which density is being reported.

Chapter 16

Porosity — The Architecture of Empty Space

Pore structure determines both durability and strength.

Empty space inside a ceramic is never neutral. It always influences behaviour. A ceramic body may appear dense from the outside, but internally it contains pores formed during shaping, drying, and firing. These pores are not simply accidental defects. They are part of the microstructure, and their size, shape, and distribution strongly influence performance.

Porosity refers to the fraction of a ceramic's volume that consists of empty space. However, total porosity alone does not tell the full story. Two ceramics may have the same overall porosity and still behave very differently. The reason is that pores differ not only in amount, but also in size, shape, and connectivity.

For this reason, porosity in ceramics is usually understood in two main forms: open porosity and closed porosity.

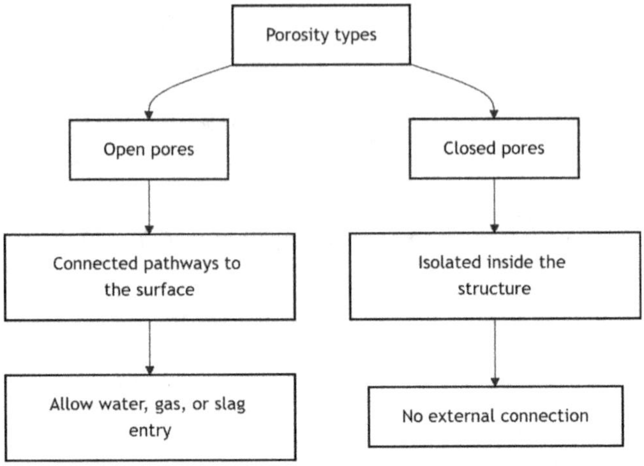

Figure 16.1 — Open vs Closed Porosity
Open pores connect to the external surface while closed pores remain isolated within the ceramic structure.

This distinction is structurally important.

Open porosity forms interconnected channels. When a ceramic is immersed in water and absorbs moisture, it is the open pore network that fills.

Closed porosity consists of isolated voids trapped within the structure. These pores lower density, but they do not allow fluids to enter the body.

Because of this, open and closed pores do not affect service behaviour in the same way.

If open porosity is high, water absorption increases, permeability rises, and chemical penetration becomes easier. In floor tiles, this increases the risk of damage under freezing conditions because absorbed water expands on freezing. In sanitaryware, open porosity reduces hygiene and stain resistance. In refractories, open pores permit molten slag or metal to penetrate the structure, accelerating degradation.

Closed porosity behaves differently. It still lowers strength because it reduces the effective load-bearing area, but it can improve thermal insulation because trapped air reduces heat transfer. This is why porosity cannot be judged simply as high or low. It must always be understood by type and function.

Porosity is also closely linked to permeability. Permeability describes the ability of gases or liquids to move through the material. It depends not only on how much pore volume exists, but on whether pores are connected.

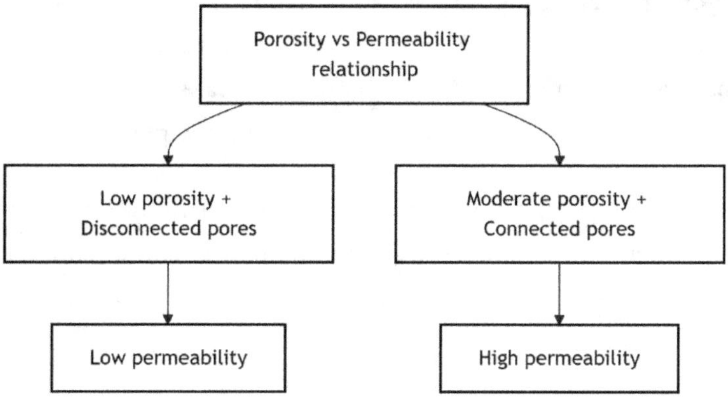

Figure 16.2 — Porosity vs Permeability Relationship
Increasing open porosity increases the pathways available for fluid flow through the ceramic body.

This relationship explains why two materials with similar porosity may show very different service performance. A ceramic with moderate porosity but isolated pores may remain relatively impermeable. Another with similar porosity but connected pores may allow rapid penetration.

In high-temperature refractories, this distinction is critical. Permeability strongly affects corrosion resistance because connected pores allow slag and reactive species to move into the lining. Even a small increase in connected porosity can shorten service life. In filter ceramics, high permeability is desirable. In dense structural ceramics, it is not.

Pore size also matters. Large pores behave as major stress concentrators and act as likely crack initiation sites under tensile loading. Fine, uniformly distributed pores reduce strength less severely because each individual stress concentration is smaller. In some thermal shock applications, fine pores may slightly improve crack resistance by interrupting crack propagation. However, excessive porosity of any kind reduces mechanical reliability.

Porosity is strongly influenced by processing. During forming, particle packing determines the initial void structure. A good particle size distribution helps fine particles fill the spaces between coarse ones, reducing large voids. During firing, sintering reduces porosity as particles bond and the structure densifies. Insufficient firing leaves excessive open pores. Excessive firing may trap gases and create bloating. Drying also influences porosity, because rapid moisture removal can generate microcracks that later behave as connected pores.

Porosity therefore affects many properties at once. It influences strength, water absorption, thermal conductivity, permeability, chemical resistance, creep resistance, and long-term durability. It connects directly to density, microstructure, and fracture behaviour.

Engineering Meaning of Porosity

Table 16.1 — Porosity and Its Performance Impact

Porosity Characteristic	Structural Meaning	Engineering Effect
High open porosity	Connected pore network	High absorption and permeability
High closed porosity	Isolated voids	Lower strength, better insulation
Large pore size	Major stress concentrators	Reduced bending strength
Fine uniform pores	Controlled microstructure	Balanced strength and insulation
Low total porosity	Dense structure	High strength, low permeability
Non-uniform porosity	Local weak zones	Crack initiation risk

This table summarizes the engineering meaning of pore architecture.

In structural tiles, open porosity must usually be minimized. In insulating refractories, porosity is deliberately engineered. In chemical process environments, permeability must be tightly controlled because connected pores allow fluid penetration. These differences reflect the same ceramic principle: internal structure governs service behaviour.

Ceramics are flaw-controlled materials, and pores act as structural flaws. Because ceramics do not yield before fracture, pore presence and connectivity strongly influence durability. When pores connect, penetration risk increases. When pore size increases, strength decreases. When porosity is controlled carefully, performance becomes more predictable.

Porosity is therefore not just empty space. It is the architecture of the void system inside the ceramic. That architecture determines how stress moves, how heat flows, how fluids penetrate, and ultimately how long the material survives.

The next chapter connects porosity and density directly to mechanical strength, showing why load capacity in ceramics depends so strongly on the elimination and control of voids.

◈ Industrial Example 8 — Porosity Balance in Steel Ladle Refractories

Situation:
Refractory bricks used in steel ladles must withstand rapid temperature changes and aggressive slag environments.

Initial assumption:
Higher density and lower porosity would always improve refractory performance.

Reality:
Moderate porosity can improve thermal shock resistance by helping the structure absorb thermal stress.

Correction:

Design refractory compositions with controlled porosity that balances thermal shock resistance with chemical durability.

Lesson:

Porosity in refractories must be optimized, not simply minimized.

Explanation:

Pores can help relieve thermal stress by interrupting crack propagation and reducing the driving force for fracture during temperature changes. This improves thermal shock resistance. However, excessive porosity allows molten slag and metal to penetrate the refractory structure. Once penetration occurs, chemical reactions weaken grain boundaries and accelerate material degradation. Refractory design therefore balances porosity to maintain both thermal stability and resistance to chemical attack.

Decision Insight:

- Open porosity increases permeability and penetration risk.
- Closed porosity lowers strength but improves insulation.
- Water absorption indicates connected pore networks.

Chapter 17

Density–Strength Link — How Porosity Controls Strength

Uniform compaction is the foundation of predictable performance.

Strength in ceramics is not created. It is preserved by eliminating voids.

In the previous chapters, density and porosity were defined separately. Their full meaning becomes clear when they are connected directly to mechanical strength. Ceramics do not fail because the average material is weak. They fail because stress concentrates at flaws, and pores are among the most important flaws within the structure.

Each pore reduces the effective load-bearing area of the ceramic. At the same time, it creates a local stress concentration. For this reason, density and strength are closely linked. As porosity increases, strength decreases. However, this reduction is not simple or proportional. If a body contains 10% pore volume, strength does not fall by only 10%. The reduction is usually much greater because stress is amplified around the pore boundaries.

Under tensile loading, cracks often initiate at pores. Large pores behave almost like pre-existing cracks. Sharp or irregular pore shapes intensify stress concentration further. For this reason, porosity becomes one of the most critical structural parameters governing both strength and reliability.

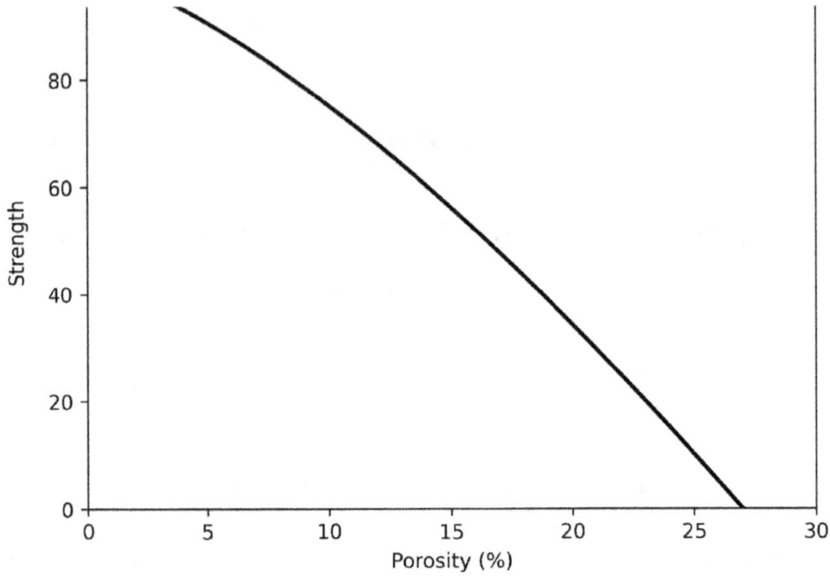

Figure 17.1 — Porosity vs Strength Trend
As porosity increases, the effective load-bearing area decreases and the strength of the ceramic material declines.

This trend explains many practical observations in ceramics.

At very low porosity, stress passes through a relatively continuous solid network and strength remains high. As porosity increases slightly, strength decreases gradually. Beyond a certain threshold, however, strength falls sharply because the solid framework becomes less continuous and the influence of pore-related stress concentration becomes dominant.

This is why small increases in porosity can sometimes produce large drops in strength. For example, a rise in open porosity from about 3% to 6% may reduce strength noticeably but still remain manageable. A rise from around 10% to 15%, however, may cause a much larger decline. At higher porosity levels, even small structural changes can have major mechanical consequences.

Pore size is just as important as pore volume. Large pores behave like large flaws and may dominate fracture even when the total porosity is only moderate. Fine, evenly distributed pores reduce strength less severely because the local stress concentration around each pore is lower. Uniform pore structure improves reliability. Non-uniform pore structure increases variability.

Density variation within a single component is especially dangerous. A region of lower density becomes a weak zone where cracks are more likely to initiate under bending or thermal stress. In floor tiles, density uniformity across thickness helps prevent edge cracking and warpage. In refractory bricks, density strongly affects both compressive strength and creep resistance.

In general, higher density improves room-temperature bending and compressive strength because the body contains fewer pores and a more continuous solid framework. However, excessive densification combined with high glassy phase formation may reduce hot strength. As in other ceramic systems, structural balance remains important.

The density–strength relationship is strongly shaped by processing. Proper particle size distribution improves packing. Higher compaction pressure increases green density. Controlled firing reduces residual porosity. Stable cooling limits microcrack formation. Every stage influences final strength through its effect on pore structure.

Strength in ceramics is therefore not an isolated property. It is the mechanical expression of internal structure. If density is low, strength cannot be high. If porosity is high, stress concentration increases. If pore distribution is uneven, failure becomes less predictable. Larger components also carry greater risk because the probability of larger flaws increases with volume.

These observations connect directly to the ten laws of ceramic behaviour established earlier in the book. Ceramics are flaw-controlled. Pores act as structural flaws. Flaws amplify stress. Ceramics do not yield before fracture. For this reason, density and porosity directly govern fracture behaviour and load capacity.

Engineering Meaning of Density–Strength Relationship

Table 17.1 — Porosity and Mechanical Behaviour

Structural Condition	Mechanical Consequence	Practical Meaning
Very low porosity	High bending strength	Suitable for load-bearing use
Moderate porosity	Moderate strength	Controlled structural use
High porosity	Low strength	Insulation-focused design
Large isolated pores	Severe local stress	Early crack initiation
Fine uniform pores	Improved reliability	Balanced performance
Density gradient	Weak zones	Unpredictable fracture

This table gives the practical interpretation.

If bending strength falls below specification, the first investigation should often focus on density and pore distribution. If fracture surfaces show large cavities, forming and firing conditions should be reviewed. If strength variability is high, density uniformity within the component must be examined.

Ceramic strength is not improved by adding something to the surface after the structure is already flawed. It is achieved by refining the internal structure itself. Ceramics do not gain reliability through external correction. Strength is preserved by minimizing voids and controlling the architecture of porosity within the body.

With this understanding, the discussion can now move into Part V — Raw Materials, where ceramic behaviour begins at the mineral level. Before porosity forms, particles must exist. Before particles behave, minerals determine their structure. In ceramics, strength begins long before firing. It begins with raw materials.

Decision Insight:

- Density variation produces strength variation.
- Uniform compaction improves structural reliability.
- Monitor density distribution across the component.

End of Part IV — Density & Porosity

Density defines structure.
Porosity defines pathways.
Uniformity defines predictability.
Variations in compaction are variations in performance.

Part V — Raw Materials Intelligence

Understanding the Building Blocks of Ceramic Bodies

The behaviour of a ceramic product begins long before forming or firing. Raw materials determine plasticity, phase development, thermal expansion, and final microstructure.

This part explains the roles of clay minerals, quartz, feldspar, particle size distribution, and material purity.

Understanding these relationships allows engineers to design bodies that behave predictably during processing and service.

Chapter 18

Clay and Plasticity — Why Ceramics Can Be Shaped

Shaping behaviour begins at the particle level.

The strength of a ceramic begins long before firing. Before a ceramic body can resist load as a rigid material, it must first be shaped while it is still soft and deformable. Every ceramic product begins as a collection of mineral particles. These particles must be brought together, rearranged, and formed into shape before firing transforms them into a hard and brittle structure.

This shaping ability exists because of clay. Clay gives ceramic bodies the capacity to deform under pressure without cracking and to retain their new shape after the pressure is removed. Without this behaviour, ceramic manufacturing would not be possible. Bricks could not be extruded, sanitaryware could not be cast, tiles could not be pressed, and refractories could not be moulded.

This behaviour does not arise from water alone. It develops through the interaction between clay mineral structure and water. Among the clay minerals used in traditional ceramics, kaolinite is one of the most important because it helps explain how this mechanism arises.

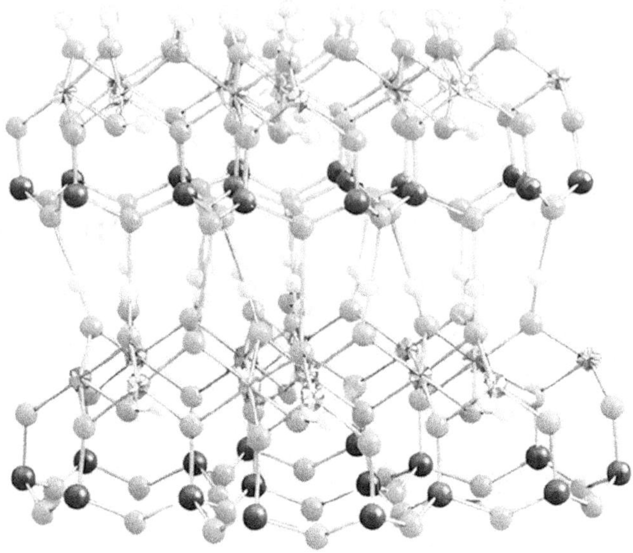

Figure 18.1 — Layered Crystal Structure of Kaolinite

Kaolinite consists of stacked silicate layers that enable plastic deformation during forming.

Kaolinite consists of repeating 1:1 layers made of a silica tetrahedral sheet and an alumina octahedral sheet. These sheets bond together to form stable layers, while adjacent layers are held by hydrogen bonding. This produces the non-expanding layered structure characteristic of kaolinite and generates the fine plate-like particles that strongly influence ceramic forming behaviour.

This layered structure is important because shaping behaviour begins with particle shape. Clay particles are extremely fine and plate-like. When dry, these particles are friction-locked. They resist movement and the body behaves in a brittle way. Once water is added, thin films forms around the particles. These films reduce friction and allow particles to slide past one another when pressure is applied.

Deformation becomes possible because the particles can rearrange while still maintaining enough contact to keep the body together. When forming pressure stops, electrostatic attraction and mechanical interlocking help the body retain its new shape.

A. Dry Clay Powder (No Water)

```
+--------------------------------+
| [] [] []  [] []   |
|   []  [] []   []  |
| [] []   [] []     |
+--------------------------------+
```

- o Clay particles loosely packed
- o High interparticle friction
- o No particle mobility
- o Body behaves brittle

B. Water Added (Water Films Form)

```
+--------------------------------+
| []~  []~  []~   []~  []~  |
|   []~   []~  []~    []~   |
| []~  []~    []~  []~      |
+--------------------------------+
```

- o ~ = thin water film surrounding particles
- o Water forms lubricating films
- o Friction between particles decreases
- o Particles become mobile

C. Plastic Deformation During Forming

```
+----------------------------------+
| []~ []~ []~ []~ []~ []~      |
|   []~ []~ []~ []~            |
| []~ []~ []~ []~ []~          |
+----------------------------------+
```

- o Particles slide past each other
- o Structure rearranges under pressure
- o Clay body deforms without cracking
- o Plastic shaping becomes possible

Figure 18.2 — Mechanism of Plasticity Development in Clay Bodies
Water layers between clay particles allow particle sliding, producing plastic behaviour during shaping.

In dry clay, particles are loosely packed but friction-locked, so the body behaves in a brittle manner. When water is added, thin films forms around the particles and reduce friction. Under pressure, the particles slide and rearrange, allowing the body to deform without cracking. This lubricated particle movement gives clay its characteristic forming behaviour.

This behaviour depends on several factors acting together. Particle size, particle shape, surface chemistry, and water content all influence how easily particles move. Fine, plate-like particles increase surface interaction and promote sliding. Water controls mobility. If too little water is present, particles cannot rearrange and the body cracks during shaping. If too much water is present, cohesion is lost and the body becomes unstable.

Useful forming behaviour therefore exists only within a controlled range.

Different clays behave differently. Kaolin is relatively coarse and pure, so it provides moderate deformability with lower drying shrinkage. Ball clay is much finer and more workable, but it also increases drying shrinkage. In industrial bodies such as tiles, clay content must therefore be balanced carefully. Too little clay reduces formability, while too much clay increases shrinkage, warpage, and cracking risk.

Clay behaviour also affects forming density. A body with sufficient workability compacts more uniformly and forms more consistently. However, excessive clay content can trap air, encourage lamination, or create instability during shaping. In extrusion, low workability can produce edge cracking. In pressing, uneven moisture distribution can create density gradients that later appear as warpage or fracture.

Clay behaviour also has a major influence on drying. As water leaves the body, clay-rich systems shrink because the particles move closer together. If shrinkage is uneven, internal tensile stress develops. Microcracks may form during drying and remain hidden until firing or service reveals them as weakness.

Clay therefore enables shaping, but it also introduces structural risk. It must be controlled, not maximized. Clay minerals do more than assist shaping. During firing they decompose and form new phases that become part of the final ceramic structure. Uneven clay distribution therefore causes more than uneven forming; it can also lead to uneven shrinkage, uneven phase development, and internal stress after firing.

Clay is not merely a shaping aid. It is the starting structural foundation of the ceramic body.

Engineering Meaning of Plasticity

Table 18.1 — Plasticity and Processing Behaviour

Factor	Effect on Plasticity	Industrial Impact
Fine particle size	Increases plasticity	Easier forming
Plate-like structure	Enables sliding	Shape retention
Optimal water content	Balanced mobility	Stable shaping
Excess water	Weak cohesion	Slumping, cracking
High ball clay content	High plasticity	Higher drying shrinkage
Low clay content	Low plasticity	Forming defects
Controlled plasticity	Balanced structure	Reduced defect risk

This summary shows the practical meaning of clay behaviour in ceramic processing. Workability makes shaping possible, but excessive clay content increases shrinkage and cracking risk, while insufficient clay reduces forming stability.

Clay behaviour therefore connects directly to the process–property–performance chain. Raw materials define potential. Forming behaviour controls shaping. Shaping determines density, and density ultimately influences strength.

For this reason, clay selection and moisture control are not minor adjustments. They are strategic structural decisions. Ceramics begin as soft systems governed by water and mineral structure. After firing they become rigid and brittle, but their future strength and reliability are rooted in this early stage.

The next chapter examines quartz and feldspar, which balance clay behaviour, control shrinkage, and govern bonding during firing. Clay alone cannot produce a stable ceramic body. It must be balanced.

In ceramics, softness begins the journey, but structure determines the end.

❖ Industrial Example 9 — Black Core in Porcelain Tiles

Situation:
A dark core was visible in the cross-section of fired porcelain tiles.

Initial assumption:
The discoloration was suspected to result from raw material contamination.

Reality:
Organic materials inside the body did not fully oxidize because oxygen supply during firing was insufficient.

Correction:
Extend the oxidation stage in the firing cycle before densification begins.

Lesson:
Chemical reactions must be completed before vitrification closes the structure.

Explanation:
During early firing, organic matter and carbonaceous compounds must oxidize and escape as gases. If the body begins to densify before this oxidation is complete, the developing vitrified structure restricts oxygen entry and gas escape. Unburned carbon then remains trapped in the interior, producing the characteristic black core. Adequate oxygen availability and sufficient oxidation time before densification are therefore essential.

Decision Insight:

- Excess plasticity increases drying shrinkage and cracking risk.
- Moisture content must be controlled precisely.
- Raw material variation affects forming stability.

Chapter 19

Quartz and Feldspar — The Balance of Stability and Bonding

Composition determines expansion, vitrification, and structural balance.

A ceramic body survives not because of one ingredient, but because of balance. A ceramic body is not a random blend of minerals. It is a controlled combination of materials, each contributing a different function during forming, drying, firing, and cooling. If clay alone is shaped and fired, the body becomes unstable. Drying shrinkage becomes excessive, warpage increases, cracking risk rises, and the final strength may remain limited.

To control this behaviour, non-plastic materials such as quartz and feldspar are introduced. Each has a different role in stabilizing the body. Clay provides plasticity during shaping and contributes alumina to the fired structure. Quartz reduces excessive shrinkage and improves dimensional stability. Feldspar acts as a flux and forms a glassy phase during firing, bonding the particles together.

The performance of the ceramic body therefore depends on how these three components are balanced. This relationship is commonly represented using a ternary diagram, which maps the compositional balance between clay, quartz, and feldspar.

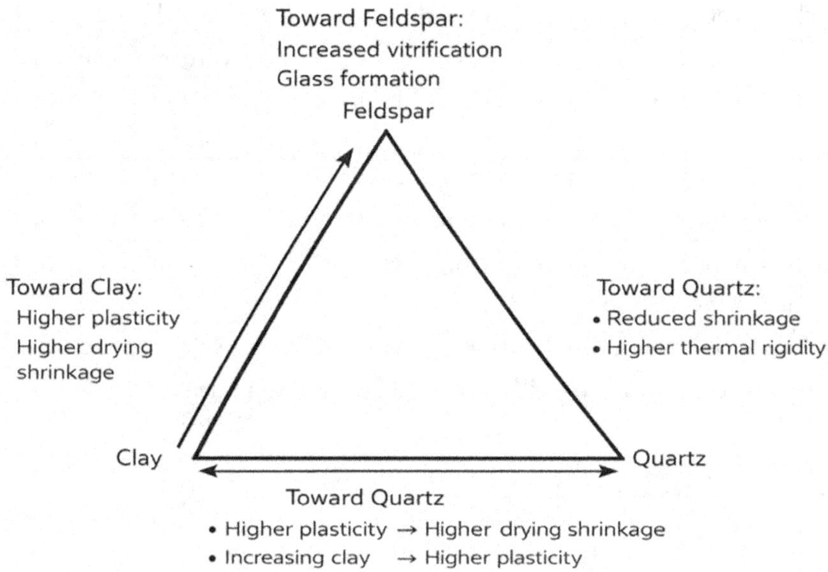

Figure 19.1 — Ternary Diagram (Clay–Quartz–Feldspar)
A ternary composition diagram illustrating the balance between clay, quartz, and feldspar in ceramic body formulations.

This diagram is more than a composition chart. It is a behaviour map.

Every ceramic body occupies a position within this triangle, and that position predicts how the body will behave during forming, drying, firing, and cooling.

When the composition shifts toward the clay corner, plasticity increases and shaping becomes easier. However, drying shrinkage also rises, and the risk of cracking during drying becomes greater.

When the composition shifts toward the quartz corner, drying shrinkage decreases and dimensional stability improves before firing. However, quartz does not contribute strongly to bonding. If quartz content becomes too high, the body weakens after firing because the bonded area between particles decreases. High quartz content can also increase thermal rigidity and make the body more sensitive to temperature change.

When the composition shifts toward the feldspar corner, vitrification increases during firing. Feldspar softens, forms a liquid phase, fills pore space, and bonds particles together. Density and strength increase as a result. However, excessive feldspar creates too much liquid phase. The body may deform at peak temperature, and a high glass content can reduce high-temperature strength and increase creep.

This balance explains many observations seen in industrial production. A body that cracks during drying often contains too much clay or too little quartz. A body that remains porous after firing often lacks sufficient feldspar. A body that deforms in the kiln usually contains excessive feldspar. The ternary diagram therefore helps diagnose problems logically. Ceramic recipe adjustment is not guesswork. It is movement within this behavioural triangle.

Functional Roles in Detail

Each component contributes a distinct structural function.

Clay — The Structural Former

Clay provides plasticity and allows the body to be shaped. During firing, clay minerals decompose and transform into alumina-rich crystalline phases that contribute to the structural framework of the fired ceramic. However, clay also shrinks significantly during drying and firing. High clay content therefore raises the risk of internal stress and cracking. Clay is essential, but it must be controlled.

Quartz — The Dimensional Regulator

Quartz is largely non-plastic and acts as a rigid filler. It reduces drying shrinkage and improves dimensional stability during shaping and drying. However, quartz does not strongly participate in bonding and remains largely crystalline after firing. If too much quartz is present, the body weakens because the bonded area between particles decreases. Quartz must therefore reduce shrinkage without weakening the structure.

Feldspar — The Bonding Agent

Feldspar functions as a flux during firing. At elevated temperatures, it softens and forms a liquid phase that moves between particles, fills voids, and later solidifies as a glassy matrix during cooling. This reduces porosity and increases strength. If feldspar is insufficient, densification remains incomplete. If it is excessive, too much liquid forms and dimensional control is lost. Feldspar therefore governs vitrification.

Balance and Firing Behaviour

During firing, these three components interact to create the final microstructure. Clay-derived phases form the crystalline framework. Quartz particles remain embedded within that framework. Feldspar melts partially and forms a glassy phase that bonds the structure together.

The final body therefore becomes a composite of three main elements: a crystalline framework derived from clay, residual quartz particles, and a glassy bonding phase formed from feldspar. The relative proportion of these phases determines strength, shrinkage, thermal expansion, creep resistance, and long-term durability.

The clay–quartz–feldspar system therefore governs both processing behaviour and final performance.

Engineering Meaning of the Clay–Quartz–Feldspar System

Table 19.1 — Functional Balance in Ceramic Bodies — Clay, Quartz and Feldspar

Component	Functional Role	Risk if Excessive	Risk if Insufficient
Clay	Plasticity and alumina source	High shrinkage, cracking	Poor formability
Quartz	Shrinkage control	Reduced bonding, thermal rigidity	Excess shrinkage
Feldspar	Glassy phase formation	Over-vitrification, deformation	Low strength, high porosity
Balanced system	Controlled densification	Stable performance	—

This table clarifies the practical meaning of compositional balance.

The objective in ceramic formulation is not to maximize one component, but to maintain the right balance between all three. This connects directly to the behaviour framework established earlier in the book. Composition controls how the body behaves during processing. Processing shapes the microstructure. Microstructure determines properties.

When composition is imbalanced, process control alone cannot fully correct the resulting problems.

Ceramic design therefore begins within this compositional triangle. Even small shifts toward one corner can change behaviour significantly. Balance creates predictability. Predictability creates reliability. Reliability defines long-term performance.

The next chapter examines particle size and purity, because even when composition is correct, behaviour still changes if particle scale and cleanliness are not controlled. In ceramics, chemistry establishes the framework, but balance determines survival.

Decision Insight:

- Quartz regulates shrinkage and thermal expansion.
- Feldspar controls vitrification and bonding.
- Balanced composition determines final performance.

Chapter 20

Particle Size and Purity — Where Structure Truly Begins

Raw material control defines final structural integrity.

Before a ceramic becomes solid, it exists only as a system of particles. At this stage, nothing is yet bonded and nothing is rigid. The future structure of the ceramic exists only as potential. That potential is shaped first by particle size and then by how those particles are distributed and how pure they are.

Two ceramic bodies may contain the same proportions of clay, quartz, and feldspar, yet behave very differently during processing and firing. One may fire to high strength with low warpage. The other may crack, deform, or remain porous. The chemistry is the same, but the particle system is different.

This is why structure truly begins at the particle stage.

When particles are all similar in size, they do not pack efficiently. Each particle touches its neighbours, but voids remain between them. These voids later become pores after firing. Pores reduce the effective load-bearing area of the body and act as stress concentrators.

When a range of particle sizes is present, smaller particles can fill the spaces between larger ones. Packing becomes more efficient, green density rises, and the initial void volume decreases. Packing is therefore the first structural decision in ceramics. Poor packing guarantees higher porosity later. Good packing reduces weakness before firing even begins.

Particle size distribution influences the entire process chain. It affects green density, drying shrinkage, firing shrinkage, final porosity, and mechanical strength.

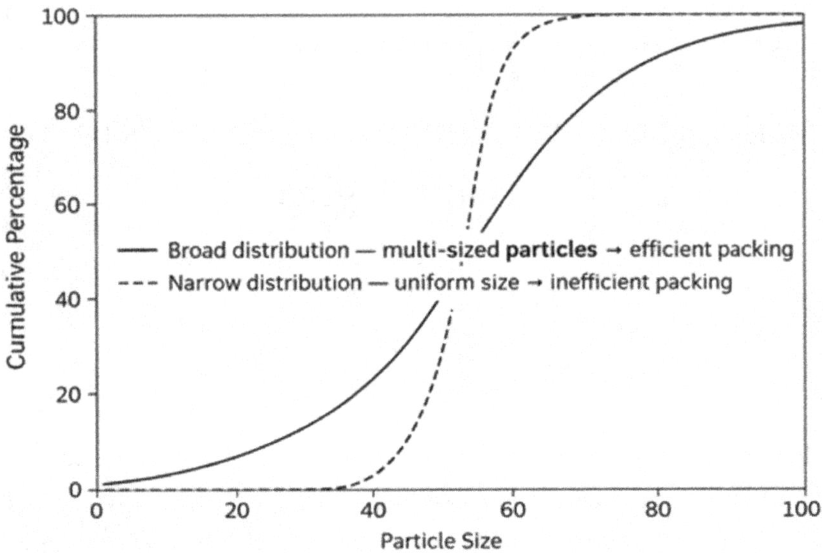

Figure 20.1 — Particle Size Distribution Curve
Graph showing the distribution of particle sizes in ceramic powders and its influence on packing and densification.

A broad particle size distribution generally gives better structural packing. Large particles form the framework, medium particles fill intermediate spaces, and fine particles occupy the smallest gaps. This layered arrangement reduces void volume and improves green density.

A narrow particle size distribution behaves differently. Because the particles are similar in size, they leave larger gaps between one another. These gaps become pores, and the final body tends to remain weaker and more porous.

Packing directly influences green density. Higher green density means particles begin closer together, so less movement is needed during sintering. Shrinkage becomes more controlled and dimensional stability improves. If packing is uneven, shrinkage also becomes uneven, creating internal stress that can lead to warpage or cracking during firing.

Particle size also affects sintering behaviour. Fine particles sinter more easily because they have higher surface area and shorter diffusion distances. They can therefore bond at lower temperatures. However, extremely fine particles also increase water demand during forming, which raises drying shrinkage and cracking risk. Coarse particles reduce water demand and shrinkage, but if they are not balanced with finer particles, they leave large voids in the structure.

Particle size also influences the pore system after firing. Coarse particles tend to leave larger residual pores, which act as more severe flaws and reduce strength. Fine, evenly distributed pores are less harmful. For this reason, particle size distribution influences not only density, but also the final flaw architecture within the ceramic.

Purity is equally important. Even when particle size distribution is optimized, impurities can destabilize the system. Iron contamination may change melting behaviour. Alkali impurities may lower softening temperature. Organic contamination may interfere with drying or firing. During firing, impurities may create unwanted phases or local glass pockets, leading to expansion mismatch, colour variation, or structural weakness.

In high-performance refractories, even small impurity levels can reduce hot strength. In white ceramics, impurities affect colour and surface quality. In structural ceramics, impurity phases may behave as crack initiation sites. Particle size governs physical packing. Purity governs chemical predictability. Together, they determine how the microstructure evolves during firing.

Once firing is complete, that microstructure is fixed. Poor particle design cannot be corrected afterward.

This is why ceramic strength is not created suddenly in the kiln. It is prepared beforehand through intelligent selection, distribution, and control of particles.

Engineering Meaning of Particle Size and Purity

Table 20.1 — Particle Characteristics and Structural Consequences

Parameter	Structural Effect	Engineering Outcome
Broad PSD	Efficient packing	Higher green density
Narrow PSD	Larger void volume	Higher final porosity
Fine particles	Faster sintering	Increased shrinkage risk
Coarse particles	Larger residual pores	Strength reduction
Uniform distribution	Even shrinkage	Dimensional stability
High purity	Controlled phase formation	Predictable performance
Impurities	Uncontrolled reactions	Variability and defects

This table summarizes the engineering meaning of particle characteristics.

Particle size distribution determines how tightly the structure begins. Purity determines how predictably that structure evolves during firing. Packing establishes density. Density influences strength. Strength governs reliability.

Ceramics are flaw-controlled materials, and pores act as structural flaws. Those pores originate from particle arrangement, which depends directly on particle size distribution. Particle design is therefore not merely a laboratory detail. It is structural engineering at the microscopic level.

In ceramics, strength is not added later. It is prepared at the particle stage.

The next chapter moves into forming behaviour, where particles are mobilized, aligned, compacted, and shaped into a coherent body. Particles alone do nothing. Their movement determines how the structure forms, and the structure ultimately determines survival.

🗳 Decision Insight:

- Particle size distribution controls packing efficiency.
- Fine particles improve packing but increase shrinkage risk.
- Raw material purity ensures predictable firing behaviour.

End of Part V — Raw Materials Intelligence

Ceramic Behaviour System begins at the particle level.
Clay, quartz, and feldspar are not ingredients.
They are structural decisions made before firing.

Part VI — Forming & Green Stage

Shaping the Structure Before the Kiln

Before firing, ceramic components exist in a fragile but critical state known as the green body. At this stage, density distribution, particle packing, and forming uniformity are established.

This section explains slip rheology, powder processing, casting behaviour, and forming defects. Many failures observed after firing originate from weaknesses introduced during these early stages.

Chapter 21

Slip Rheology — Controlling Flow to Control Structure

Flow behaviour determines internal uniformity.

Before a ceramic body develops strength, it must first develop internal order. This order begins when particles are still suspended in water. The way particles move, interact, and arrange themselves during suspension determines how they will pack during forming. That packing ultimately defines the internal structure of the ceramic.

A ceramic slip is not simply wet clay. It is a controlled suspension of mineral particles in water. Inside this suspension, particles are constantly interacting. They attract, repel, cluster, disperse, settle, and rearrange.

These interactions determine how the suspension flows. The science that describes this behaviour is rheology.

In ceramics, rheology represents more than simple flow behaviour. It controls how particles organize themselves before drying and firing fix the structure permanently. Because particles remain mobile in the suspension stage, rheology becomes the last opportunity to influence internal structure before the body becomes rigid.

Particle Interaction in Suspension

The behaviour of particles in a slip depends mainly on the balance between attractive and repulsive forces.

If attraction between particles becomes strong, they stick together and form clusters known as flocs. These clusters trap water between particles and create an open, irregular structure. The suspension becomes thick and unstable.

When repulsive forces dominate, particles remain separated and mobile. Water moves freely between particles, allowing them to rearrange easily during forming. This condition is known as deflocculation.

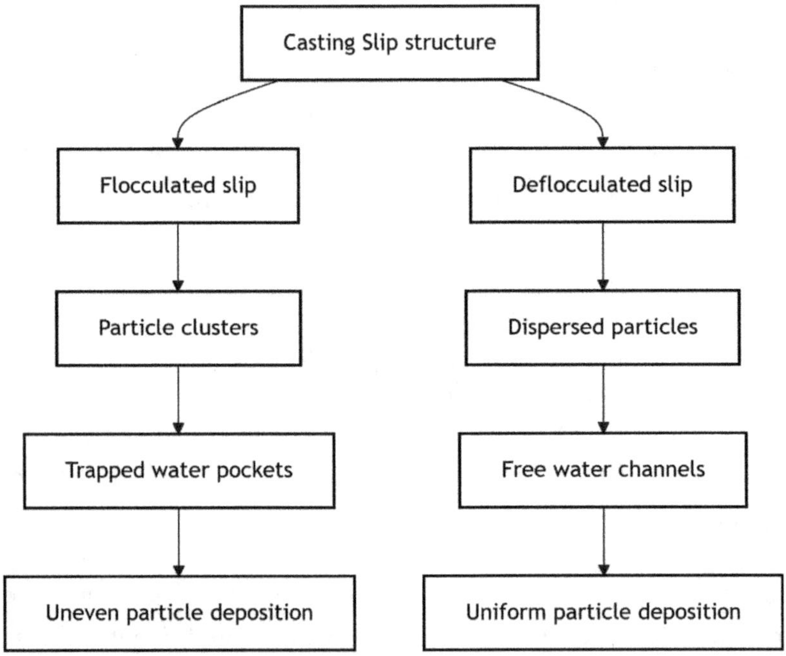

Figure 21.1 — Flocculated vs Deflocculated Structure
Comparison of particle arrangements in flocculated and deflocculated ceramic suspensions.

In a flocculated slip, particles attract each other and form clusters. These clusters trap water and produce unstable flow. During casting, the clusters deposit unevenly and create density variations.

In a deflocculated slip, particles remain dispersed and separated by electrostatic repulsion. Water can move freely between particles, improving fluidity and enabling more uniform particle packing.

Even though these changes occur at the microscopic scale, their effects are visible in the final product.

Slight flocculation can produce wall thickness variations during casting. In sanitaryware and tableware this may cause distortion. In refractory bodies it can produce weak zones. In tile bodies it can generate density gradients that later lead to warpage.

Slip chemistry controls this particle interaction. Electrolytes adjust surface charge on the particles, and deflocculants create repulsion that keeps particles separated. However, excessive deflocculant can destabilize the system again. Rheology therefore must be carefully controlled rather than simply maximized.

Solid Content and Viscosity

Another critical factor in slip rheology is solid content, also known as solid loading.

A slip with low solid content flows easily because it contains more water. However, high water content creates other problems. During drying, more water must evaporate, which increases shrinkage and raises the risk of cracking.

Increasing solid content reduces drying shrinkage and improves green density. Higher green density usually leads to lower porosity and better strength after firing.

However, increasing solids also increases viscosity. When too many particles occupy the suspension, they interfere with each other's movement and the slip becomes thick and difficult to process.

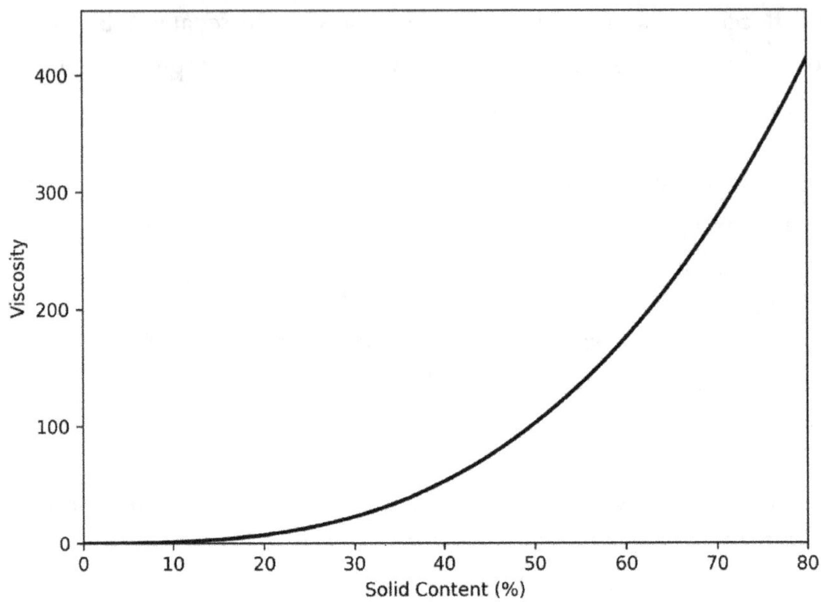

Figure 21.2 — Viscosity vs Solid Content Curve
Increasing solid content in ceramic slips raises viscosity and alters flow behaviour during forming.

At low solid content, particles move easily and viscosity remains low. As solid content increases, particle interactions grow stronger and viscosity gradually rises.

Beyond a critical concentration, particle crowding causes viscosity to increase sharply, making the suspension difficult to pump or cast.

This relationship defines the operating window of a ceramic slip.

The goal is to maintain the highest possible solid content while preserving stable flow. If solid content is too low, drying shrinkage becomes excessive. If solid content is too high, the slip becomes difficult to process and air entrapment increases.

Process Stability in Suspensions

Rheology also affects several practical processing issues.

One of these is air entrapment. If viscosity becomes too high or mixing is unstable, air bubbles can remain trapped inside the suspension. During firing these bubbles become pores within the ceramic body.

Pores act as flaws, reducing strength and reliability.

Rheology also influences sedimentation. In poorly dispersed slips, heavier particles may settle during storage. This creates density gradients inside tanks or casting reservoirs. When such slips are used, different castings may develop slightly different compositions.

Process variability then becomes structural variability, which ultimately becomes strength variability.

Another important factor is slip ageing. Over time the surface charge balance between particles may change. As a result, viscosity may slowly drift.

If this change is not monitored, casting behaviour may gradually shift as well. Stable viscosity over time usually indicates stable particle interaction, while drifting viscosity signals potential instability in the suspension.

For this reason, rheology must be measured regularly rather than assumed.

Structural Consequences of Rheology

Table 21.1 — Rheology and Structural Impact

Rheological Condition	Structural Result	Engineering Consequence
Flocculated slip	Clustered packing	Density variation
Optimal dispersion	Tight uniform packing	Higher green density
Low solid content	High drying shrinkage	Crack formation
Excess solid content	Poor flow, air entrapment	Porosity defects
Stable viscosity	Consistent casting	Predictable performance

Slip rheology determines how particles arrange themselves before water leaves the system. Once drying begins, particle positions become fixed. When firing starts, the microstructure becomes permanently locked.

Rheology therefore represents the last stage where structure can still be controlled while particles remain mobile.

Ceramics are flaw-controlled materials, and many flaws originate during the particle stage. Particles that are poorly dispersed in suspension cannot reorganize later in the process.

In ceramics, the relationship is direct:

Flow determines packing.
Packing determines density.
Density determines strength.

Strength, in turn, determines long-term reliability.

The next chapter moves to powder preparation and pressing, where particles are compacted mechanically rather than through fluid flow. Whether particles move under hydraulic flow or mechanical force, the objective remains the same: build the structure before the kiln makes it permanent.

Decision Insight:

- Viscosity controls casting behaviour.
- Particle dispersion determines packing uniformity.
- Deflocculation must be controlled, not maximized.
- Solid loading determines green density.

Chapter 22

Powder and Forming — Structure Built Under Pressure

Geometry and density are fixed before firing begins.

The kiln does not fix structure. It only preserves what forming has already created.

Before firing, before sintering, before strength appears, a ceramic body exists as a compacted arrangement of particles. In pressing-based processes, this arrangement is created mechanically. The structure that later determines strength, shrinkage, and reliability is already being built inside the powder bed.

Spray drying is the beginning of this mechanical structuring. Liquid slip is atomized into fine droplets and brought into contact with hot air. Water evaporates rapidly. Each droplet collapses into a granule made of fine particles loosely bonded by residual moisture. These granules must flow easily into the die cavity and must break and rearrange efficiently during pressing.

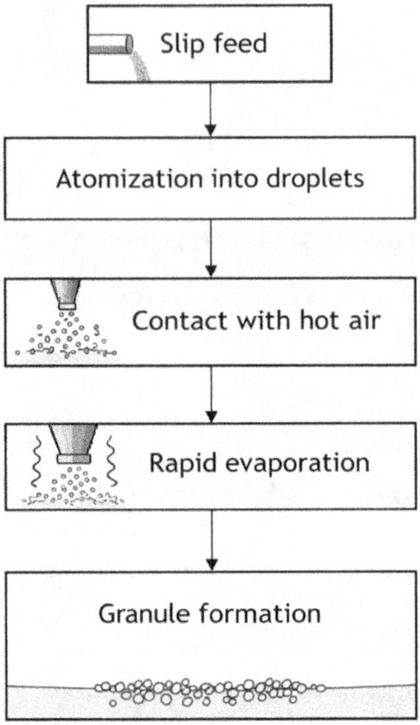

Figure 22.1 — Spray Dryer Schematic
Illustration of spray drying used to convert ceramic slurry into free-flowing granulated powder.

A spray dryer converts ceramic slurry into free-flowing granulated powder by atomizing the liquid feed into fine droplets inside a stream of hot air. Rapid evaporation removes moisture and forms spherical granules with controlled size and moisture content. These granules provide good flowability and uniform compaction during powder pressing.

Figure 22.1 represents this transformation. What leaves the spray dryer is not just powder; it is engineered granulate. Granule size, internal porosity, and moisture content determine how compaction will occur.

If granules are too coarse, die filling becomes uneven. If they are too fine, powder flow decreases and air entrapment increases. When moisture content is insufficient, particles resist rearrangement during pressing, while excessive moisture can lead to lamination and the formation of internal slip planes.

Spray drying therefore determines how particles respond to force during forming. Pressing then applies that force. As pressure is applied through the punch, granules fracture, fine particles rearrange, and air between particles is expelled. The contact area between particles increases and green density develops. However, the applied pressure does not distribute uniformly throughout the compact.

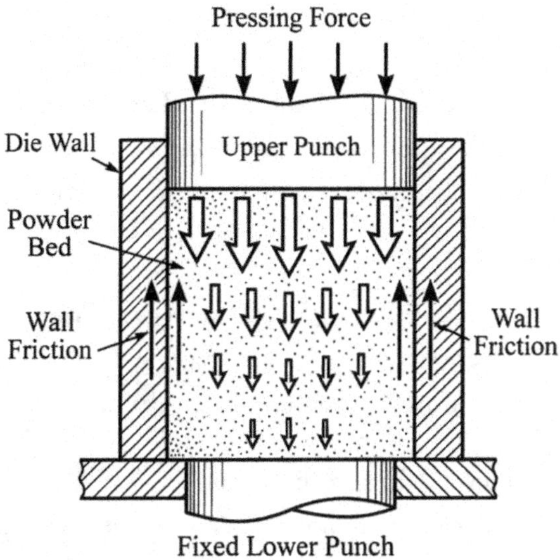

Figure 22.2 — Pressing Force Distribution
Compaction pressure during pressing creates density gradients inside the ceramic body.

During powder pressing, applied force is transmitted through the granulated ceramic powder inside the die. Friction between powder particles and the die walls causes the pressure to decrease with depth, leading to non-uniform density distribution within the compact. This variation can influence shrinkage, strength, and defect formation during firing.

Figure 22.2 illustrates the internal stress gradient created during pressing. Friction between powder and die walls absorbs part of the applied force. As a result, pressure decreases with depth. Upper layers may compact more than lower layers. Edges may differ from centre.

This density gradient becomes critical during firing. Regions with higher green density shrink less, while regions with lower density shrink more. The resulting differential shrinkage introduces bending and internal tensile stress. In many cases, warpage begins here—not in the kiln, but during pressing.

Pressing speed also influences this behaviour. If pressure increases too rapidly, air cannot escape through inter-particle channels. The trapped air forms internal lamination cracks that may remain invisible in the green body but can open significantly during firing.

Tool design also affects stress distribution. Large-format products, thick components, and complex geometries tend to amplify density variation. Friction conditions at the die wall, influenced by lubrication, further alter compaction uniformity.

Mechanical forming therefore becomes structural engineering at the particle scale. The press is not only shaping geometry; it is shaping density. Density determines shrinkage, shrinkage determines stress, and stress ultimately determines failure.

If green density is uneven, firing cannot correct it. The kiln can densify the compact further, but it cannot remove density gradients—only intensify their effects. In this sense, the structural destiny of the product is already established within the green body.

Engineering Meaning of Powder and Pressing

Table 22.1 — Forming Variables and Structural Consequences

Variable	Structural Effect	Performance Impact
Uniform granule size	Even die filling	Stable density
Incorrect moisture	Poor particle rearrangement	Lamination risk
High die wall friction	Density gradient	Warpage
Rapid pressure application	Air entrapment	Internal cracks
Controlled compaction	Uniform shrinkage	Dimensional stability

Powder forming converts loose particles into a defined internal structure. Although this structure may appear stable in the green state, it remains fragile, and both drying and firing will test its integrity.

Ceramics are flaw-controlled materials, and many flaws originate from density variations and trapped air created during pressing. During forming, structure is built under pressure. If that structure is uniform, the product can survive later processing stages. If it is uneven, failure may only be delayed.

The kiln does not repair forming mistakes; it preserves them permanently. The next stage, drying, challenges the compacted structure because the water that allowed particle movement must leave without generating stress. When forming has been imbalanced, drying reveals these weaknesses immediately.

Decision Insight:

- Granule quality controls compaction behaviour.
- Density gradients create warpage and cracking.
- Pressing uniformity defines structural reliability.

Chapter 23

Green Strength — The Strength Before Strength

Weakness introduced early survives through the kiln.

A ceramic body must first survive before it can become strong. After forming, the body has shape, density, and geometry, but it does not yet possess fired strength. At this stage it is held together only by mechanical interlocking between particles and by small capillary forces created by residual moisture. This condition is known as the green state.

Green strength is temporary and fragile, yet it determines whether the body can survive handling, drying, and entry into the kiln without internal damage. Although a green body may appear solid and feel rigid when handled, internally it is still only a network of particles pressed together by mechanical force. There are no sintering bonds, no glassy phase, and no crystallized strength—only particle contact.

Green strength mainly depends on particle packing, compaction pressure, and moisture condition. When particles are packed tightly during pressing, the contact area between them increases, frictional resistance grows, and mechanical interlocking becomes stronger.

If compaction is uneven, however, density varies across the body. Regions of lower density contain larger voids and weaker particle contacts, creating weak zones that can later develop into lamination cracks.

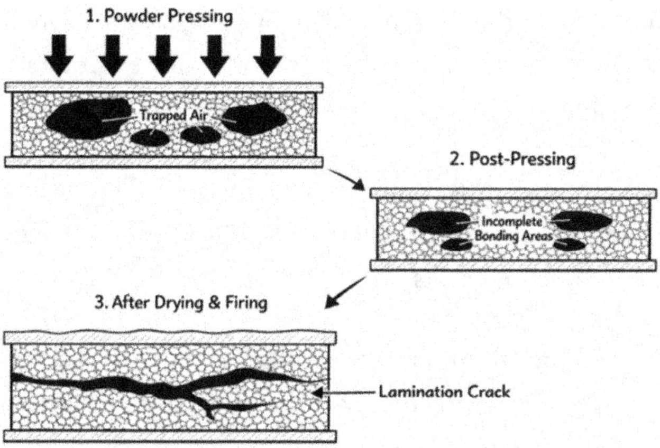

Figure 23.1 — Lamination Crack Formation
Air entrapment or poor powder flow during pressing can produce internal lamination cracks.

Figure illustrates the mechanism.

During powder pressing, the compact may form in layers. If air becomes trapped or bonding between layers is incomplete, a weak horizontal plane develops inside the compact. During drying or firing, stresses can cause this weak plane to open, producing a lamination crack.

This plane may not be visible in the green body, but during drying the shrinkage of the material generates tensile stress across that interface. During firing, thermal expansion and further shrinkage amplify this stress, eventually causing the lamination crack to open. The failure therefore was not created in the kiln; it originated earlier in the pressing stage.

Moisture also plays a dual role during compaction. A small amount of moisture helps particles slide and rearrange, allowing tighter packing under pressure. After pressing, this moisture creates capillary forces between particles, contributing to green cohesion.

If moisture content is too low, particle rearrangement remains incomplete, the contact area between particles is reduced, and green strength decreases. If moisture content is too high, internal slip planes may form during pressing, and these planes can later act as crack origins.

Pressing speed strongly influences the ability of air to escape from the powder bed. When pressure rises too quickly, trapped air cannot move through the compacting powder and remains inside as internal voids.

These voids create local regions of lower density. During drying, these regions shrink more than surrounding areas, producing shrinkage differences that generate stress and eventually open cracks.

Handling introduces an additional stress condition. Large tiles, slabs, sanitaryware, or refractory blocks may bend slightly under their own weight during lifting or transport. If green strength is marginal, microcracks can initiate even before drying begins. These cracks are often invisible, but ceramics are flaw-controlled materials, and invisible flaws remain structurally significant.

When firing begins, sintering bonds form around existing flaws. The flaw does not disappear; it becomes embedded within the final microstructure. For this reason, green strength must be uniform throughout the body.

Uniform green density leads to uniform drying behaviour, which reduces stress concentration and prevents crack initiation. Green strength is therefore not simply a matter of achieving maximum compaction, but of achieving consistent compaction across the entire body.

Engineering Meaning of Green Strength

Table 23.1 — Green Strength Variables and Structural Consequences

Variable	Structural Effect	Performance Consequence
Uniform green density	Even particle contact	Stable handling
Density gradient	Local weak zones	Crack initiation
Trapped air	Lamination planes	Delamination during firing
Optimal moisture	Capillary bonding	Improved cohesion
Excess moisture	Internal slip planes	Structural weakness
Rapid pressing	Air entrapment	Hidden defects

Green strength is temporary, but its consequences are permanent.

If lamination exists in the green body, firing will not remove it. Likewise, if density varies within the body, shrinkage during firing will vary as well. Uneven shrinkage leads to stress concentration, which can eventually result in cracking or distortion.

The green structure represents the final condition before drying begins. Once drying starts, the removal of water begins to test the internal cohesion of the body. When the green structure is uniform, drying stresses distribute more evenly. When the structure contains flaws or density variations, drying quickly reveals these weaknesses.

In ceramics, strength does not begin in the kiln. It begins with particle contact. The green stage is therefore the point where this contact network is established and where the foundation of the final structure is defined.

◈ Industrial Example 10 — Lamination Crack in Ceramic Tiles

Situation:
An internal horizontal crack became visible in the tile after firing.

Initial Assumption:
The crack was attributed to shrinkage stress during firing.

Reality:
Air trapped between powder layers during pressing created a weak internal plane within the tile body.

Correction:
Adjust powder moisture content and optimize pressing parameters to prevent air entrapment.

Lesson:
Defects originate early in the manufacturing process.

Explanation:

During powder pressing, air must escape from the compact as particles rearrange under pressure. If moisture content, powder flow, or pressing speed are not properly controlled, pockets of air can become trapped between layers of powder. These trapped air layers create planes of weakness inside the green body.

During firing, shrinkage and thermal stresses act along these weak interfaces, causing the layers to separate and form lamination cracks. Proper powder conditioning and controlled pressing allow air to escape, producing a uniform and structurally sound compact.

Decision Insight:

- Particle contact defines green strength.
- Density gradients create crack origins.
- Green structure survives the kiln.
- Uniform compaction ensures structural integrity.

End of Part VI — Forming & Green Stage

The kiln does not correct early mistakes.
Green defects become fired defects.
Uniformity at forming is the foundation of reliability.

Part VII — Drying & Firing

Transforming Powder into Structural Ceramic

Drying and firing represent the transition from fragile formed bodies to strong ceramic structures. During these stages, moisture removal, sintering, and vitrification occur.

This part explains the physics of drying, the stages of firing, and how heating and cooling rates influence internal stress, phase development, and defect formation.

Chapter 24

Drying — The First Structural Test

Moisture removal must occur without creating stress imbalance.

Drying does not create weakness. It exposes imbalance. After forming, the ceramic body contains water that enabled shaping. This water allowed particles to slide, rearrange, and compact, making forming possible. However, once the shape is established, the same water becomes a structural risk and must be removed.

Drying may appear simple because it involves evaporation, but it is actually the first stage where internal stress can develop within the body. If moisture leaves uniformly, the structure remains stable. If moisture removal becomes uneven, cracks may form.

Water in the green body exists in different forms. Free water occupies larger pores between particles and moves relatively easily. At the start of drying, surface water evaporates first, and capillary forces draw water from the interior toward the surface. During this stage, the body shrinks as particles move closer together while water volume decreases.

Once particles come into full contact and form a rigid network, further shrinkage becomes limited. However, moisture still remains in smaller pores and as adsorbed water on particle surfaces. At this stage the mechanism changes: water removal continues, but shrinkage slows. If moisture distribution becomes uneven, internal stresses begin to develop.

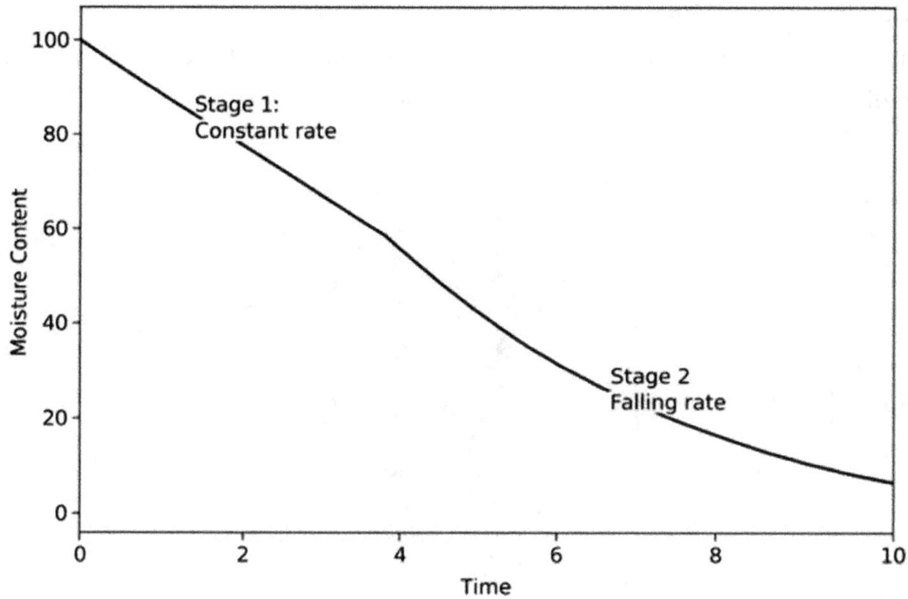

Figure 24.1 — Drying Curve (Constant & Falling Rate)
Drying proceeds through an initial constant-rate stage followed by a slower falling-rate stage as moisture becomes harder to remove.

Figure represents the drying behaviour.

In the constant-rate period of drying, evaporation at the surface controls the process. Internal moisture movement is fast enough to continuously supply water to the surface, so evaporation proceeds steadily and shrinkage occurs uniformly.

In the falling-rate period, the situation changes. Internal diffusion of moisture becomes slower than surface evaporation, causing the surface to dry faster than the interior. A moisture gradient develops within the body.

This gradient creates stress. The outer layer begins to contract while the core still contains moisture and remains swollen. As the surface attempts to shrink inward while being restrained by the wetter interior, tensile stress develops in the surface layer.

Ceramic failure is governed primarily by tensile stress, cracks tend to originate at the surface during this stage. The risk increases with thicker sections, since moisture must travel a longer distance to reach the surface.

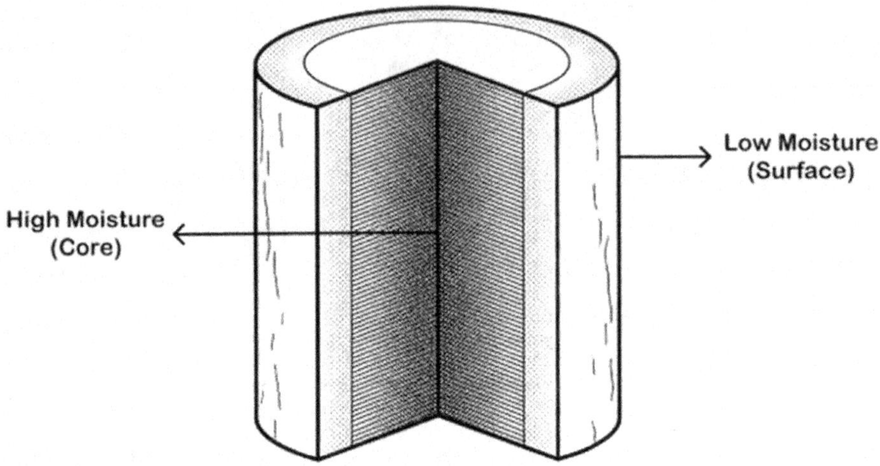

Figure 24.2 — Moisture Gradient Through Thickness
During drying, moisture gradients develop between the surface and interior, generating internal stress.

Figure illustrates this internal imbalance.

During drying, moisture is removed first from the surface while the core retains higher moisture content. The drier surface begins to shrink earlier than the interior, creating a moisture gradient through the thickness. This differential shrinkage generates tensile stress at the surface, which can lead to cracking if the stress exceeds the tensile fracture strength.

In thin ceramic products, moisture travels only a short distance to the surface, so the moisture gradient remains small and drying stresses stay low. The body can usually tolerate this contraction without damage. In thicker products such as refractory blocks, large tiles, or sanitaryware sections, moisture must travel farther. The surface dries and shrinks while the interior still contains water, creating tensile stress in the outer regions. When this stress exceeds green strength, cracking occurs. Drying cracks therefore result from restrained shrinkage as the surface contracts while the interior resists.

Drying rate strongly affects stress development. Temperature, humidity, airflow, and product geometry control how quickly moisture is removed. If surface evaporation becomes too rapid compared with internal moisture diffusion, a strong moisture gradient forms and internal stress rise quickly, causing cracks before drying is complete. Uneven airflow can worsen this effect, as edges and corners dry faster than central regions.

Density variations from forming also influence drying. Lower-density areas retain more water and shrink more, while denser regions shrink less. This differential shrinkage creates bending forces that can lead to warpage. Plasticity and particle size also play a role, since fine and highly plastic bodies retain more water and slow moisture diffusion, increasing gradient risk.

Drying therefore reflects earlier processing decisions. Particle size affects water retention, forming determines density uniformity, and initial moisture content controls total shrinkage. Many cracks that appear after firing actually originate during drying as microcracks. For this reason, drying must be carefully controlled. The objective is uniform moisture removal rather than maximum speed, using balanced temperature, airflow, and humidity to minimize tensile stress and protect structural integrity.

Engineering Meaning of Drying Control

Table 24.1 — Drying Variables and Structural Consequences

Variable	Structural Effect	Performance Consequence
Rapid surface evaporation	Large moisture gradient	Surface cracking
Thick cross-section	Slow internal diffusion	Internal stress
Density variation	Uneven shrinkage	Warpage
High plasticity	High contraction	Crack risk
Controlled drying schedule	Balanced stress	Structural stability

Drying is the first stage at which the ceramic body must resist internal stress without the benefit of sintering bonds. At this point, particles are held together only by mechanical interlocking and limited cohesion from residual moisture. No strong crystalline bonds or glassy phases have formed to reinforce the structure. Although the body may appear stable externally, it remains internally fragile as water begins to leave the system.

Green strength alone must carry the load during drying. As moisture evaporates, the body contracts and tensile stresses develop. If forming produced uniform particle packing and consistent density, shrinkage occurs evenly and the body can tolerate the contraction without cracking.

However, if forming introduced density gradients, these become critical during drying. Lower-density regions shrink more because they contain greater pore space and more removable water, while higher-density regions shrink less. This difference in shrinkage creates internal restraint, and restraint generates tensile stress.

Drying therefore reveals structural imbalance rather than correcting it. Water originally enabled particle movement and compaction during shaping, but once geometry is fixed, its removal becomes a test of whether the particle network can contract safely.

As drying continues, particles move closer together and eventually form a rigid network. At this point shrinkage slows and the body becomes dimensionally stable. Although it now feels stronger, true mechanical strength has not yet developed because sintering has not occurred. The body remains sensitive to impact and thermal variation.

The next stage is firing, where chemical and physical transformations create permanent bonds between particles. Clay minerals decompose, new phases form, and liquid phases promote densification. These changes gradually convert the fragile particle network into a bonded microstructure capable of carrying load.

Once firing begins, however, the internal structure can no longer adjust. The particle arrangement, density distribution, and flaws present after drying become embedded in the microstructure. The kiln cannot correct earlier imbalance—it only strengthens the structure that already exists. What was built during forming and revealed during drying becomes permanent during firing.

◈ Industrial Example 11 — Drying Cracks in Sanitaryware Production

Situation:
Cracks appeared in sanitaryware products after firing, even though no visible defects were observed during forming.

Initial assumption:
The cracks were caused by firing defects or handling damage.

Reality:
Uneven drying during the green stage created internal tensile stress inside the body.

Correction:
Control drying conditions by maintaining proper humidity and ensuring gradual, uniform moisture removal.

Lesson:
Drying must be controlled carefully to prevent internal stress development in the green body.

Explanation:
When the surface of a ceramic body dries faster than the interior, the outer layer begins to shrink while the inner region still contains moisture. The interior resists this shrinkage, creating tensile stress within the body. Because green ceramics have limited strength, this stress can produce microcracks during drying. These cracks may remain invisible until firing, when thermal stresses open them further and make them visible. Controlled drying allows moisture to leave the body more uniformly, reducing internal stress and preventing cracking.

Decision Insight:

- Moisture gradients create tensile stress.
- Surface shrinkage begins before the core.
- Uniform drying prevents cracking.

Chapter 25

Firing Fundamentals — When Heat Makes Structure Permanent

Heat transforms particles into structure and locks in microstructure.

Firing does not create quality. It permanently records the quality you built before. After drying, the ceramic body is rigid but weak. It holds shape because particles are pressed together, not because they are bonded. The structure is temporary. It can still be damaged easily. It has not yet become ceramic in the true structural sense.

Firing changes that.

Inside the kiln, heat activates transformation. Chemical reactions occur. Particle boundaries begin to diffuse. New phases form. Glass may appear. Pores shrink. Bonds grow.

Once this transformation is complete, the structure cannot rearrange. It becomes permanent.

This is why firing must be understood as a structural transformation process, not simply a heating cycle.

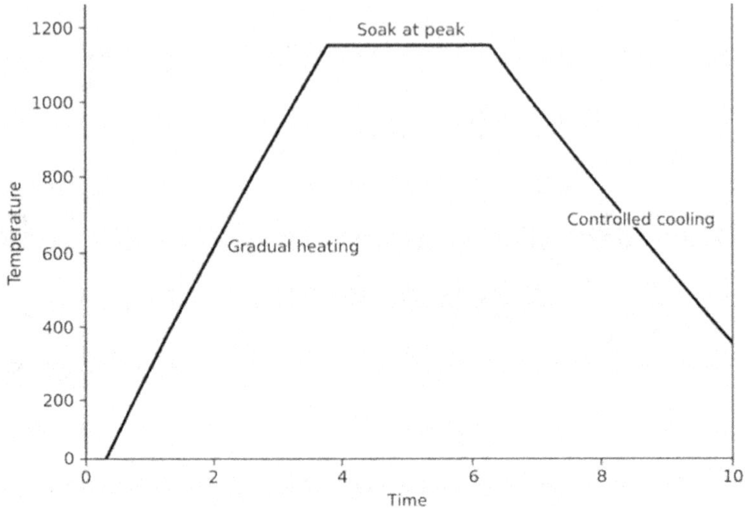

Figure 25.1 — Typical Kiln Firing Curve
Temperature versus time profile showing heating, soaking, and controlled cooling stages in ceramic firing.

Figure shows the thermal journey.

A kiln firing cycle consists of three main stages. The temperature first rises gradually during the heating stage to avoid thermal shock and allow proper reactions. At the peak temperature, the material is held for a short soak period to complete sintering and phase transformations. After soaking, the kiln is cooled in a controlled manner to prevent thermal stress and cracking.

The first stage of firing removes residual physical moisture remaining in the ceramic body after drying. Even when a piece appears dry, small amounts of water remain within fine pores and along particle surfaces. This moisture must be removed gradually as temperature rises in the kiln. If heating is too rapid, the remaining water may vaporize suddenly.

Rapid steam formation within the pore structure creates internal pressure. In open pores this pressure may dissipate, but in partially closed pores vapor cannot escape easily. Pressure builds and may generate microcracks or internal damage before sintering begins. These early defects are often invisible but weaken the structure.

As temperature rises further, chemical changes begin in the mineral structure. Clay minerals start to decompose, releasing chemically bound water from the crystal lattice. At the same time, organic materials such as binders or natural organic matter burn away. These reactions produce gases that must diffuse through the pore network and escape from the body.

If heating is too aggressive, gas evolution may occur faster than gases can escape. As temperature increases and densification begins, the pore network becomes more restricted. If gases are still being generated during this stage, they may become trapped inside the structure.

Trapped gases can expand at higher temperatures and produce defects such as bloating, blistering, or internal swelling. These defects reduce density and weaken the final product.

For this reason, the heating ramp must match the transformation behaviour of the material. Heating through moisture removal, burnout, and decomposition stages must allow sufficient time for vapours and gases to escape before densification begins.

As temperature continues to rise, particle interaction becomes stronger. Atomic mobility increases, allowing atoms to diffuse across particle surfaces and grain boundaries. At contact points between neighbouring particles, small bridges of material form as atoms migrate toward these regions.

This marks the beginning of sintering. The loose particle assembly gradually transforms into a bonded structure as contact points expand into necks. As these necks grow, pores shrink and density increases.

Sintering therefore converts a fragile compacted particle network into a continuous microstructure capable of carrying load. Once this stage progresses, the particle arrangement created during forming and drying becomes locked into place, and the bonds formed during sintering give the ceramic body its final structural strength.

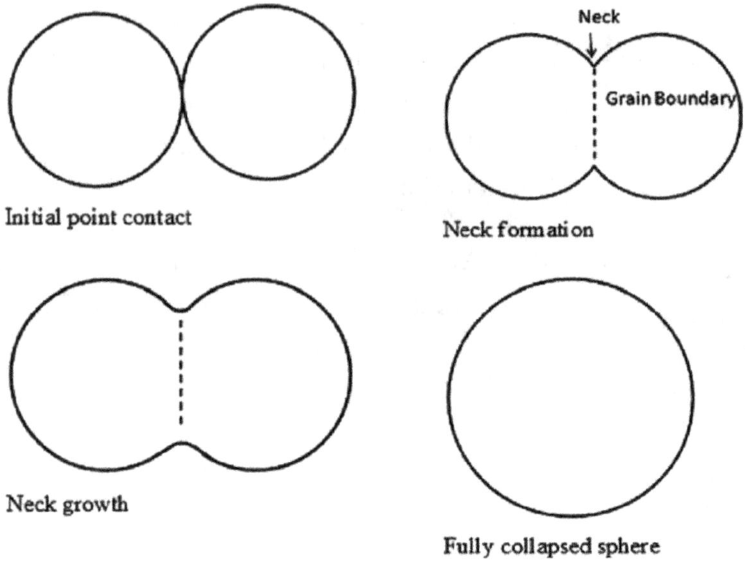

Figure 25.2 — Sintering Neck Growth
Adjacent ceramic particles bond during firing as diffusion forms necks between particles.

Figure represents what happens at the microscopic level.

During sintering, individual particles first make contact at small points. As temperature increases, material diffusion begins and a small neck form between particles. Continued diffusion enlarges the neck, strengthening the particle bond. At later stages, pores shrink and the structure gradually densifies.

At high temperature, atoms within ceramic particles gain thermal energy and become mobile. They migrate along particle surfaces and grain boundaries from regions of higher surface energy to lower energy locations, particularly at contact points between neighbouring particles.

At these contacts, small material bridges form between particles. These bridges are called sintering necks**.** With increasing temperature and time, atomic diffusion enlarges these necks, strengthening the bonds between particles.

As necks grow, particles move slightly closer together and pore spaces shrink. This reduction in pore volume increases the overall density of the body. The shrinkage observed during firing is therefore the macroscopic result of microscopic neck growth and pore elimination.

Larger necks create stronger particle bonds and increase resistance to crack initiation and propagation. Mechanical strength therefore rises as sintering progresses.

However, sintering cannot correct structural variations already present in the green body. Density gradients created during forming remain during firing. Regions with higher green density shrink less, while lower-density regions shrink more. Because these regions are connected, differential shrinkage creates internal constraint, generating bending forces and internal stress that may lead to distortion or cracking. In this way, firing often magnifies structural imbalance originating in forming.

Peak firing temperature determines the extent of sintering. If temperature is too low, diffusion remains limited and densification is incomplete, leaving residual porosity and reduced strength. If temperature is too high, excessive liquid phase may form, particularly in bodies containing fluxes such as feldspar. This can soften the structure and cause deformation or abnormal grain growth, reducing fracture resistance.

Time at peak temperature also influences the final microstructure. Short soaking may leave incomplete densification, while excessive soaking may cause grain coarsening or excessive vitrification.

Temperature uniformity within the kiln is critical. Uneven heating causes different regions to densify at different rates, producing internal stress that can result in warpage or cracking.

Cooling is also a structural stage of firing. As temperature decreases, the body contracts. If cooling is too rapid, the surface contracts faster than the interior, producing tensile stress that may cause cracking or residual stress. Some ceramics also undergo phase transformations during cooling that involve volume changes, which can generate additional stress if temperature passes through these ranges too quickly.

Firing therefore consists of three structural zones:

- Transformation during heating — moisture removal, burnout, and mineral decomposition
- Bond formation at peak temperature — sintering and densification
- Stress development during cooling — contraction and phase transformation

Firing ultimately reveals the structural history of the body. Particle size-controlled packing during forming. Packing determined green density. Green density defined the shrinkage path during sintering. Drying may already have introduced microcracks.

The kiln does not remove these earlier effects. It locks them permanently into the microstructure.

Firing therefore does not erase the past history of the ceramic body. It fixes that history permanently within the structure.

Structural Interpretation of Firing

Table 25.1 — Firing Control and Structural Consequences

Firing Variable	Structural Effect	Performance Outcome
Rapid early heating	Vapor pressure buildup	Cracking
Incomplete burnout	Trapped gases	Bloating defects
Optimal peak temperature	Balanced neck growth	High strength
Excess temperature	Over-vitrification	Deformation
Insufficient soak	Incomplete densification	Low reliability
Rapid cooling	Thermal gradient	Residual stress
Controlled cooling	Stress relaxation	Dimensional stability

Firing transforms a loose particle assembly into a bonded microstructure. It is the irreversible stage of ceramic processing. The kiln does not create quality by itself; rather, it reveals whether the earlier stages of processing were properly balanced.

Ceramics are history-dependent materials. The final strength measured in the laboratory reflects the combined effects of particle design, forming uniformity, drying control, and thermal management. Each step contributes to the structure that ultimately develops during firing.

Once firing is complete, the microstructure can no longer be adjusted. It becomes permanent, and in ceramics this permanence represents both strength and consequence.

Decision Insight:

- Firing does not fix poor forming.
- Sintering locks in structure.
- Burnout must complete before densification.

Chapter 26

Heating & Cooling Effects — When Temperature Turns into Stress

Thermal control determines structural stability.

Ceramics do not fear high temperature; they fear uneven temperature. A ceramic body inside a kiln does not experience temperature as a simple number but as movement—expansion, contraction, and transformation. When this movement occurs uniformly, the body remains stable. When it occurs unevenly, internal stress develops, and in ceramics such stress is rarely harmless.

Heating and cooling are therefore not passive transport stages. They are mechanical events driven by thermal change. Each degree of temperature variation produces dimensional change, and when different regions of the body expand or contract at different rates, internal forces arise. These forces do not require any external load—the kiln itself effectively becomes the loading system.

Heating — Expansion Under Constraint

During heating, the outer surface of a ceramic body is exposed to heat first and absorbs energy more quickly than the interior. The interior warms more slowly because heat must conduct inward through the material. This delay creates a temperature gradient between the surface and the core. As temperature rises, the hotter surface regions attempt to expand while the cooler interior resists that expansion.

This restraint generates tensile stress in the heated surface layer. At low temperatures, before sintering begins, the ceramic body is still fragile. Particle bonds remain weak and overall strength is low, so even modest tensile stress can open microcracks. The faster the heating rate, the steeper the temperature gradient between the surface and the core. In thin tiles the gradient equalizes quickly, but in thick refractories the gradient persists longer and the resulting stress becomes more significant.

Heating also interacts with internal reactions occurring within the ceramic body. As organic materials burn out, gases are released. If heating proceeds too rapidly, gas generation may exceed the rate at which gases can escape through the pore network. Internal pressure then builds inside the body, and microcracks may form before densification begins.

Later, as sintering starts and pores begin to close, trapped gases can no longer escape easily, which may produce bloating defects. Another critical region occurs during phase transformation temperature ranges. When certain minerals change crystal structure, they may expand or contract abruptly. If the body passes through these regions too quickly, local stress spikes can develop. For this reason, the heating rate is not simply a productivity parameter; it is fundamentally a stress control parameter.

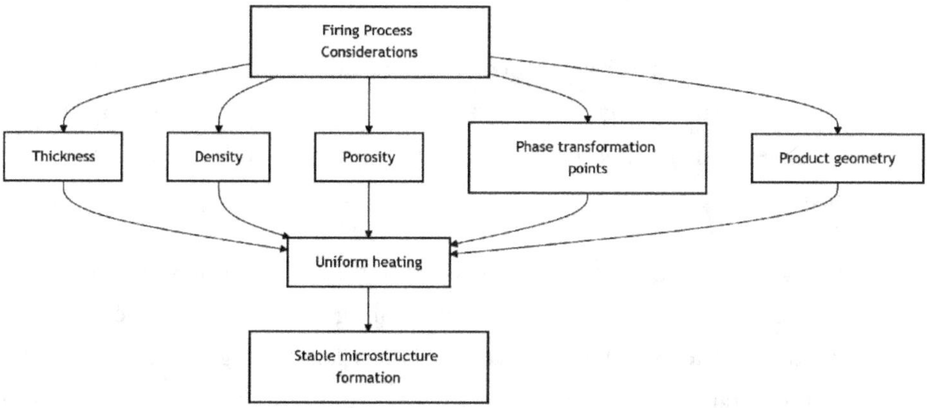

Figure 26.1 — Key Factors influencing Firing Behaviour

Key firing variables such as heating rate, temperature uniformity, atmosphere, and soak time influence final microstructure.

During kiln firing, several material and product parameters influence the thermal behaviour of ceramics. Thickness, density, porosity, phase transformation temperatures, and product geometry determine how heat is absorbed and distributed within the body. Proper control of these factors allows uniform heating, which minimizes internal stress and prevents structural imbalance before sintering bonds become permanent.

Cooling — Contraction and Locked Stress

Cooling is often more dangerous than heating. By the time cooling begins, the ceramic body has already sintered and gained strength, but it has also become brittle. As temperature decreases, the body begins to contract.

If the surface cools faster than the interior, it contracts first while the interior remains hotter and resists this movement. This restraint places the surface under tensile stress. Because ceramics are weak in tension, cracks often initiate at the surface under these conditions.

During cooling after firing, ceramics pass through temperature ranges where structural changes and thermal mismatch can generate stress. Phase transformations such as quartz inversion and differences in contraction between body and glaze create critical stress zones.

If cooling is too rapid through these temperature ranges, tensile stresses may exceed the material's strength and initiate cracking. Controlled cooling through critical temperature regions reduces thermal gradients and minimizes residual stress formation.

The cooling rate strongly influences the temperature gradient between the surface and the interior of a ceramic body. When cooling occurs rapidly, the difference between surface and core temperatures becomes larger. Edges and corners cool faster than central regions, which is why cracks often initiate at these locations.

Large-format products amplify this effect. Thick sections create stronger temperature gradients, while material characteristics also influence heat transfer.

Dense bodies conduct heat differently from porous ones, and the presence of glassy phases can change thermal diffusivity.

Temperature

↑

1200°C ————————————

Fired structure formed

Cooling begins

↓ Cooling direction

573°C ————————————

Quartz inversion zone

Sudden volume change risk

400°C ————————————

Glaze solidification

Body–glaze stress develops

200°C ————————————

Continued contraction

Room Temperature ----

Final structure locked in

Fast Cooling → High thermal stress

Controlled Cooling → Reduced residual stress

Figure 26.2 — Critical Cooling Zones in Ceramic Bodies
Certain temperature ranges during cooling are particularly sensitive to thermal stress and phase changes.

Cooling must also pass carefully through phase transformation temperatures. When contraction occurs at the same time as structural transformation, local stress concentration can increase significantly.

Residual stresses formed during cooling remain locked inside the material. Even if cracks do not appear immediately, these stresses reduce the safety margin of the ceramic. Later thermal cycling or mechanical loading may activate the stored stress. This is why some tiles crack months after installation—the origin of the failure often lies in residual stresses created during cooling rather than in the service load itself.

Rate Balance — Not Too Fast, Not Too Slow

Industrial production demands productivity, and faster firing cycles help reduce cost. However, aggressive heating and cooling narrow the safety margin for the ceramic body. Controlled rate does not necessarily mean slow firing; it means matching the heating and cooling schedule to the behaviour of the material.

Uniform temperature distribution across kiln zones is essential. Poor airflow, uneven stacking, or overloaded kiln cars can create local hot or cold spots. The ceramic body does not respond to the kiln's average temperature; it responds to its local temperature environment. For this reason, uniformity is more important than speed.

Thermal Gradient and Thickness Relationship

The magnitude of thermal stress depends on several factors, including the coefficient of thermal expansion, elastic modulus, temperature difference between regions, and structural constraint. Thicker bodies develop larger temperature differences between the surface and the core, which increases the potential for internal stress.

Materials with a high elastic modulus resist deformation more strongly, so the same temperature gradient produces higher internal stress. This is why dense porcelain behaves differently from porous earthenware during cooling. Material properties determine how much stress develops, while the heating and cooling schedule determines how large the temperature gradient becomes. Failure occurs when the resulting stress exceeds the local tensile strength.

Long-Term Impact of Thermal Control

Residual stress created during cooling influences crack resistance, glaze fit stability, thermal shock behaviour, and durability under cyclic loading. A product may pass laboratory strength tests yet still contain internal stress close to its tensile limit.

When the product later experiences temperature variation during service, the additional stress may be sufficient to initiate fracture. Heating and cooling therefore are not only manufacturing considerations; they directly influence long-term service reliability.

Structural Meaning of Thermal Rate Control

Table 26.1 — Heating & Cooling Control and Structural Outcome

Condition	Internal Effect	Structural Consequence
Rapid surface heating	Expansion mismatch	Early microcracks
Aggressive burnout	Gas pressure buildup	Bloating defects
Uneven kiln temperature	Differential shrinkage	Warpage
Rapid cooling	Surface tensile stress	Edge cracking
Fast cooling through transformation	Volume mismatch	Hidden microcracks
Uniform controlled rates	Balanced contraction	Long-term reliability

Heating introduces expansion under constraint, while cooling introduces contraction under constraint. In both cases, the dimensional change created by temperature cannot occur freely when different regions of the body respond at different rates.

This restraint converts temperature change into internal stress. In brittle materials such as ceramics, stress is not easily relieved through deformation. Instead, accumulated stress interacts with the inherent brittleness of the material and can ultimately lead to fracture.

Ceramics can tolerate extremely high temperatures, but they do not tolerate uneven temperature combined with restraint. When different regions of the body expand or contract differently, internal forces develop that act as mechanical loads within the structure.

For this reason, the kiln must be understood not only as a heating device but as a mechanical loading environment where thermal gradients continuously generate stress.

Every firing curve is a stress curve in disguise.

The heating and cooling schedule determines how quickly temperature gradients form and how much stress develops within the body. Recognizing this principle prepares us for the next structural challenge in ceramic systems: the introduction of a second material layer, the glaze, bonded to the body.

Once two materials with different thermal expansion behaviours are attached together, heating and cooling no longer act on a single structure. Instead, they act on a combined system. In such systems, differences in expansion behaviour become magnified, and the interaction between layers can amplify internal mismatch stresses within the ceramic body.

Decision Insight:

- Temperature gradients create internal stress.
- Expansion and contraction under restraint cause cracking.
- Cooling often governs residual stress.

End of Part VII — Drying & Firing

Moisture removal and sintering transform fragile shapes into structural bodies. Every temperature rise locks in structure.

Every cooling step locks in stress.

Part VIII — Glaze

Surface Engineering and Thermal Compatibility

Glazes provide surface finish, chemical protection, and aesthetic appearance. However, glaze layers must remain compatible with the ceramic body to avoid cracking, peeling, or surface defects.

This section examines glaze composition, glaze fit, surface tension effects, and the root causes of defects such as crazing, crawling, pinholes, and blisters.

Chapter 27

Glaze Fit — Agreement Under Restraint

Surface stability depends on expansion compatibility.

A glaze survives not because it is beautiful, but because it is in controlled compression. When glaze and body leave the kiln, they are no longer two separate materials. They are a bonded system that must behave as one.

At peak temperature, both are expanded. There is no disagreement yet. Stress is minimal because everything is hot and relaxed. The real test begins during cooling. Cooling forces contraction. Contraction under bonding creates restraint. Restraint creates stress.

Glaze fit is the management of that stress.

It is not decoration science.
It is structural compatibility.

The Glaze as a Thin Brittle Layer

A glaze is essentially a glass layer. Glass is stiff. Glass is strong in compression. Glass is weak in tension.

The glaze is thin compared to the body. That thickness difference is critical. The body acts as the structural backbone. The glaze is the surface skin. Because the glaze is thin, it cannot redistribute stress by plastic deformation. It behaves elastically until it cracks. Once cracked, the crack does not stop easily.

Thus, even small tensile strain is dangerous.

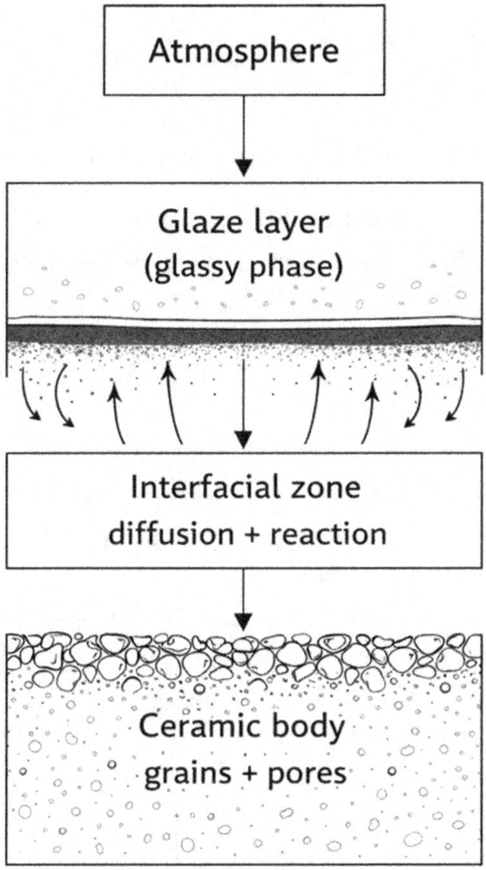

Figure 27.1 — Glaze Layer Structure
Cross-sectional view showing the glaze layer bonded to the ceramic body.

A ceramic glaze forms a glassy surface layer during firing. The glaze is exposed to the furnace atmosphere and bonds to the ceramic body through an interfacial reaction zone. This transition region develops during firing and ensures adhesion between the glaze layer and the underlying ceramic body.

Figure 27.1 shows that the glaze is fully bonded to the body. The interface is not flexible. It is fused.

That means any contraction mismatch is directly transferred across the interface. There is no cushion. If the glaze wants to contract more than the body allows, it will be stretched. If the body wants to contract more than the glaze allows, the glaze will be compressed. In both cases, force develops instantly during cooling.

Thermal Expansion — The Core Variable

Each material has a coefficient of thermal expansion (CTE). This value defines how much it changes length per degree of temperature change.

Even small differences in CTE, when multiplied by hundreds of degrees of cooling, become significant dimensional differences.

If the glaze has higher CTE than the body, it tries to shrink more during cooling. Because it is attached, it cannot shrink freely. It becomes stretched. Stretching produces tensile stress. Tensile stress produces cracks.

If the glaze has lower CTE than the body, the body tries to shrink more. The glaze resists. The glaze becomes compressed.

Compression in glass is beneficial up to a limit. Excess compression causes edge lifting and shivering.

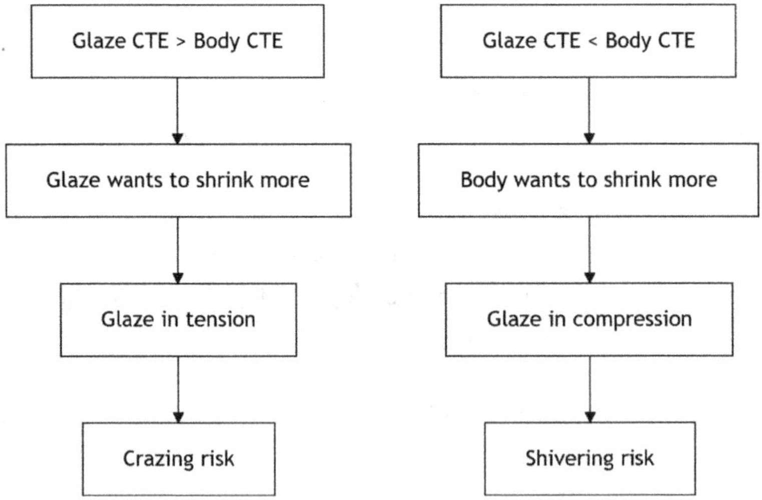

Figure 27.2 — Glaze Fit Stress Model
Thermal expansion mismatch between glaze and body generates residual stress after cooling.

During cooling after firing, glaze and ceramic body contract according to their thermal expansion coefficients (CTE). If the glaze has a higher CTE than the body, it attempts to shrink more and develops tensile stress, which can cause crazing. If the glaze has a lower CTE than the body, compressive stress develops in the glaze, which may lead to shivering or glaze peeling.

The direction of mismatch determines the type of defect. The magnitude of mismatch determines severity.

Why Slight Compression Is Desired

Slight compressive stress in the glaze helps keep microcracks closed and improves surface durability. Many glass systems intentionally use surface compression for this reason. However, compression must remain moderate. Excessive compression creates shear stress at edges and corners, leading to shivering. Glaze fit therefore aims to create a controlled and beneficial stress state.

Interaction with Body Composition

Body composition strongly influences glaze fit. Quartz increases thermal expansion, while feldspar and glassy phases modify expansion differently. Even if the glaze remains unchanged, body composition changes can shift the stress state. For example, higher quartz content increases expansion and may move the glaze from compression toward tension, increasing crazing risk.

Cooling Rate Influence

Cooling rate controls how much stress remains locked in the structure. Rapid cooling traps residual stress, while slower cooling allows partial stress relaxation. Products with identical compositions may behave differently if cooling conditions differ.

Thickness and Stress Energy

Glaze stress increases with thickness. Thick tensile layers crack more easily, while thinner layers tolerate small expansion mismatches better. Uneven glaze thickness also causes uneven stress distribution.

Delayed Failure Mechanisms

Crazing may appear long after firing. Moisture penetration and repeated temperature changes gradually extend microcracks. This delayed failure often reflects the release of stored tensile stress.

Edge and Geometry Sensitivity

Edges and corners concentrate stress and cool faster, making them common crack initiation sites.

Glaze Fit as an Engineering Decision

Glaze fit must consider body composition, glaze chemistry, firing temperature, cooling schedule, glaze thickness, and service conditions. It is a compatibility design that determines long-term surface stability.

Structural Meaning of Glaze Fit

Table 27.1 — Glaze Fit Conditions and Structural Consequences

Condition	Stress State	Visible Effect	Long-Term Behaviour
Mild compression	Controlled compressive stress	Stable surface	High durability
High compression	Edge shear stress	Shivering	Edge failure
Mild tension	Micro tensile stress	Fine crazing	Moisture-assisted growth
High tension	Strong tensile stress	Crack network	Surface degradation
Rapid cooling	High locked stress	Crack activation	Reduced life
Balanced system	Controlled compression	Stable interface	Long service life

Glaze fit is the agreement between two contraction behaviours under restraint. If they agree within tolerance, the system remains stable. If they disagree beyond tolerance, the surface reveals the truth. Crazing is not cosmetic. Shivering is not accidental. They are mechanical responses to thermal disagreement.

The glaze must not only adhere. It must contract with discipline.

Ceramics fail where stress exceeds tolerance.

Glaze fit defines whether the surface survives the cooling journey.

Next, we move to surface defects — where glaze chemistry, viscosity, contamination, and firing behaviour create visible patterns that reflect internal physics.

Because in ceramics, the surface is always telling the structural story.

◈ Industrial Example 12 — Glaze Crazing After Installation

Situation:
A fine network of cracks appeared on the glaze surface three months after installation.

Initial Assumption:
The glaze formulation was suspected to be defective.

Reality:
The ceramic body had a higher coefficient of thermal expansion (CTE) than the glaze, placing the glaze under tensile stress during cooling.

Correction:
Adjust the thermal expansion of the body or modify the glaze formulation to achieve proper glaze–body fit.

Lesson:
Glaze fit governs surface integrity.

Explanation:

Glaze and body expand and contract at different rates as temperature changes. During cooling after firing, if the body contracts more than the glaze, tensile stress develops in the glaze layer. Because glassy materials are weak in tension, this stress gradually produces a network of fine cracks known as crazing.

These cracks may not appear immediately after firing but can develop over time as environmental moisture penetrates the glaze and slightly expands the crack network.

Proper control of glaze–body thermal expansion compatibility ensures the glaze remains under slight compression, preventing crazing and maintaining surface durability.

Decision Insight:

- Slight compressive stress stabilizes glaze.
- Tensile stress produces crazing.
- Expansion mismatch defines glaze fit.

Chapter 28

Surface Defects — When the Surface Exposes the Imbalance

Surface defects reveal hidden process imbalance.

A ceramic surface does not fail quietly. It records every imbalance beneath it. The surface of a ceramic product is the most exposed and the most honest region of the structure. It is thin, brittle, and directly bonded to the body. It cannot absorb stress. It cannot yield. It cannot hide internal imbalance. When something in composition, expansion, firing timing, or gas evolution is wrong, the surface reacts first.

Surface defects are not aesthetic accidents. They are mechanical and chemical consequences made visible.

A smooth glaze layer represents a balanced interaction between melt viscosity, adhesion, gas release, cooling contraction, and thermal expansion compatibility. When one of these variables shifts outside tolerance, the surface reveals the shift immediately.

Crazing — Tension Released Across the Surface

Crazing is a network of fine cracks that develops in the glaze layer. It is a direct mechanical response to tensile stress. When glaze contracts more than the body during cooling, it is stretched. Because glass has very low tolerance for tensile strain, it fractures to release stored energy.

The cracks form in patterns that reflect stress distribution. They often initiate at edges, corners, thickness variations, or microscopic flaws where stress concentration is highest. Once initiated, cracks propagate across the surface until the tensile energy is sufficiently relieved.

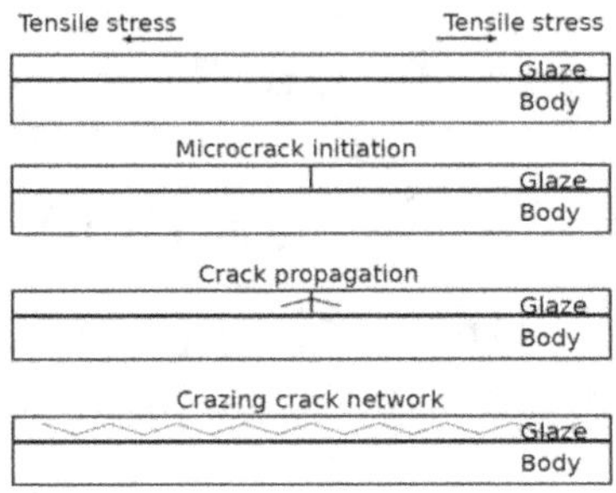

Figure 28.1 — Crazing Crack Network
Crazing appears as a network of fine cracks when the glaze is under tensile stress.

Figure represents the stress relief process. The crack network is not random decoration. It is a mechanical equilibrium pattern. Each crack reduces local tensile stress.

When a glaze layer is under tensile stress, small microcracks begin to initiate at weak points in the glassy structure. These cracks propagate through the glaze as stresses continue to act. Over time, multiple intersecting cracks form a visible crazing network across the glaze surface.

Crazing may appear immediately after firing, or it may develop gradually during service. Moisture plays a role in delayed crazing. Water penetrates microcracks and assists crack growth at the tip. Over time, fine invisible cracks become visible networks.

Crazing compromises hygiene, chemical resistance, and long-term durability. It is not merely cosmetic. It indicates that the glaze is operating in tension.

Crawling — Failure of Wetting During Melt Flow

Crawling appears as bare patches where glaze has retracted during firing. Unlike crazing, crawling is not a cooling problem. It originates during melting.

When glaze becomes molten, it must wet and adhere to the ceramic body. Wetting depends on surface cleanliness, glaze composition, and surface tension balance.

If the molten glaze has high surface tension or poor adhesion to the substrate, it pulls back instead of spreading uniformly.

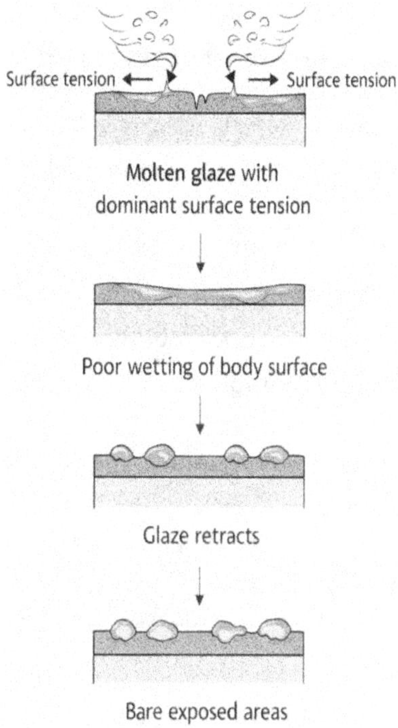

Figure 28.2 — Crawling Mechanism
Poor wetting or contamination can cause glaze to retract during firing, exposing the ceramic body.

When molten glaze has high surface tension and poor wetting ability on the ceramic body, it cannot spread uniformly. The glaze pulls back during firing, retracting into thicker islands. This retraction exposes portions of the ceramic body surface, producing the crawling defect.

Figure shows the contraction of molten glaze into islands. Crawling is often promoted by contamination such as dust, oil, or uneven drying residues on the body surface. Excessively thick glaze layers increase internal contraction forces during melting, encouraging retraction.

High viscosity at peak temperature reduces levelling ability. If glaze does not flow adequately before stiffening, surface discontinuities remain.

Crawling reflects imbalance in adhesion and melt flow. It is a firing-stage defect, not a thermal expansion mismatch.

Pinholes and Blisters — Gas Interaction with Molten Glass

Pinholes and blisters are gas-related surface defects. They result from interaction between gas evolution inside the body and glaze melting timing.

During heating, organic materials burn out and gases are released. If glaze becomes molten before gas release is complete, gas bubbles rise through the molten layer. If viscosity is high, bubbles cannot escape smoothly. When they burst at the surface, small craters remain. These are pinholes.

If glaze seals the surface early, gas becomes trapped beneath the layer. Pressure builds until a raised blister forms.

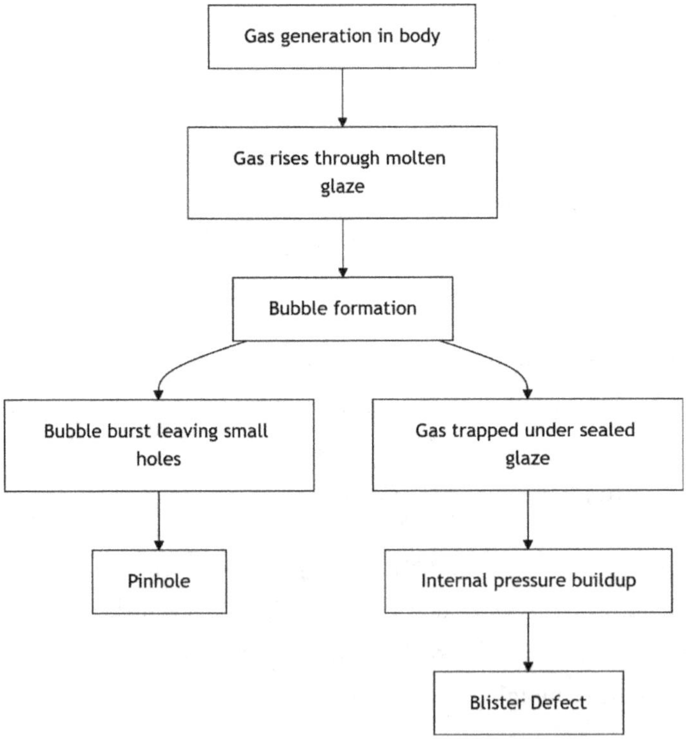

Figure 28.3 — Pinholes & Blister Formation
Gas release during firing can create pinholes or blisters in the glaze surface.

Figure distinguishes between open gas escape and trapped gas pressure.

During firing, gases generated within the ceramic body may rise through the molten glaze layer. If the gas bubble reaches the surface and bursts, it leaves a small opening known as a pinhole. If the glaze seals before the gas escapes, the trapped gas builds internal pressure and produces a blister defect in the glaze surface.

Pinholes are surface-level interruptions where bubbles break through. Blisters are structural cavities beneath the glaze, often larger and more disruptive.

Both defects depend on synchronization between burnout completion and glaze melting stage. If burnout is incomplete when glaze viscosity decreases, the probability of defect formation increases.

Thus, firing curve coordination is critical.

Surface as a Structural Indicator

The surface does not create defects independently. It reacts to imbalance.

Crazing indicates tensile stress due to expansion mismatch.
Crawling indicates poor wetting or excessive surface tension.
Pinholes indicate viscosity and gas timing mismatch.
Blisters indicate trapped gas and early glaze sealing.

Surface defects often trace back to upstream stages: raw material selection, particle size distribution, drying uniformity, glaze thickness control, and firing rate.

The surface is the final messenger of earlier decisions.

Structural Interpretation of Surface Defects

Table 28.1 — Surface Defects and Structural Origin

Defect	Immediate Cause	Deeper Structural Origin
Crazing	Tensile stress in glaze	Thermal expansion mismatch
Shivering	Excess compressive stress	Overcorrection in glaze fit
Crawling	Poor wetting	Surface contamination or high surface tension
Pinholes	Gas escape through melt	Burnout–melt timing imbalance
Blisters	Trapped gas pressure	Early glaze sealing
Uneven gloss	Incomplete levelling	Viscosity or temperature variation

Surface defects are not minor irregularities. They are the visible translation of physics into appearance.

The glaze layer is thin and brittle. It responds immediately to stress and imbalance. Because it cannot redistribute load, it fractures or retracts.

A disciplined ceramic process treats every surface defect as a diagnostic signal. The solution is not to mask the surface. The solution is to correct the structural imbalance that produced it.

The surface always tells the truth about the internal state of the system.

From here, the narrative advances into refractories, where ceramics face continuous temperature, load, and chemical attack. The lessons remain the same — stress, time, and compatibility govern behaviour.

Only the intensity changes.

❖ Industrial Example 13 — Pinholes in Glaze Surface

Situation:
Persistent pinholes appeared on the glaze surface even after glaze formulation adjustments.

Initial Assumption:
The defect was believed to be caused by improper glaze viscosity.

Reality:
Organic materials within the body did not completely burn out before the glaze began to seal the surface.

Correction:
Modify the firing schedule to allow sufficient time for gas release before glaze maturation.

Lesson:
Firing stages interact and must be properly balanced.

Explanation:
During firing, organic materials and chemically bound compounds inside the ceramic body decompose and release gases. If the glaze layer begins to melt and seal the surface before these gases have fully escaped, the gases become trapped beneath the glaze.

As the glaze continues to melt and later solidifies, these trapped gases create small openings on the surface known as pinholes. Proper firing schedules ensure that burnout reactions occur earlier in the firing cycle, allowing gases to escape before the glaze reaches its sealing stage.

❖ Industrial Example 14 — Glaze Crawling Defect

Situation:
Bare patches appeared where the glaze pulled away from the ceramic surface after firing.

Initial Assumption:
The defect was attributed to improper glaze application thickness.

Reality:
Dust contamination on the green body surface prevented proper wetting and adhesion of the glaze.

Correction:
Improve surface cleaning and remove dust or loose particles before glazing.

Lesson:
Surface preparation governs coating adhesion.

Explanation:
For glaze to adhere properly, it must wet the ceramic surface uniformly during application. If the green body surface contains dust, powder residues, or loose particles, the glaze slurry cannot spread evenly across the surface. These contaminated areas disrupt the bond between glaze and body.

During firing, as the glaze melts and begins to flow, surface tension forces pull the glaze away from these poorly bonded regions, leaving exposed patches known as crawling defects. Maintaining clean surfaces and controlled handling before glazing ensures proper glaze wetting and adhesion.

◈ Industrial Example 15 — Glaze Blistering After Firing

Situation:
Raised bubbles appeared on the glaze surface after firing.

Initial Assumption:
The defect was suspected to be caused by impurities in the glaze raw materials.

Reality:
Gas release from the ceramic body occurred after the glaze had already softened and begun to seal the surface.

Correction:

Adjust the firing profile so that gas-generating reactions occur before the glaze reaches its melting stage.

Lesson:

Timing within the firing cycle is critical.

Explanation:

During firing, several reactions inside the ceramic body release gases, including decomposition of carbonates, burnout of organic matter, and release of trapped air. If these gases evolve after the glaze has already softened and formed a viscous layer, the gases cannot escape easily. Instead, they push against the molten glaze and create bubbles.

When the glaze later solidifies, these bubbles remain trapped as raised blisters on the surface. Proper firing schedules ensure that gas-evolving reactions are completed before glaze melting, allowing gases to escape without disturbing the glaze layer.

Decision Insight:

- Surface defects reveal internal imbalance.
- Gas release must finish before glaze sealing.
- Wetting controls glaze adhesion.

End of Part VIII — Glaze

Surface is not decoration.
It is stress interaction between glass and body.
Compatibility determines durability.

Part IX — Refractories

Ceramic Performance Under Extreme Conditions

Refractories operate under the most severe environments encountered in ceramic systems. High temperatures, mechanical load, chemical attack, and thermal cycling challenge structural stability.

This part explains refractory materials, lining design, anchor systems, creep behaviour, and chemical degradation mechanisms such as slag and alkali attack.

Chapter 29

Refractories — Structure Under Continuous Assault

High-temperature performance depends on structural and chemical compatibility.

A refractory does not collapse in a day. It degrades under heat, load, and chemistry until structure can no longer resist. Refractories are the most stressed ceramics in industry. They operate where temperatures are not occasional peaks but continuous realities. Inside steel ladles, rotary kilns, cement preheaters, glass furnaces, and incinerators, refractories stand between molten material and structural steel. Their job is not simply to survive heat. Their job is to maintain geometry, contain reaction, and resist chemical dissolution under load.

Unlike conventional ceramic products, refractories live in a dynamic environment. Temperature fluctuates. Slag composition changes. Mechanical loads vary. Shutdowns introduce thermal cycling. Every one of these variables interacts with structure.

Refractory failure is rarely dramatic at first. It is progressive. Thickness reduces slowly. Cracks appear near anchors. Corners spall. Slag lines erode. Distortion begins. Only when cumulative damage exceeds tolerance does collapse become visible.

To understand refractory behaviour, we must analyse it as a structural system under continuous assault.

The Refractory Lining as a Thermal Structure

A furnace lining is not a single material. It is layered engineering.

The hot face is exposed to flame or molten material. Behind it lies an insulating layer. Outside that sits the steel shell. Between refractory and shell, anchors or expansion allowances control constraint.

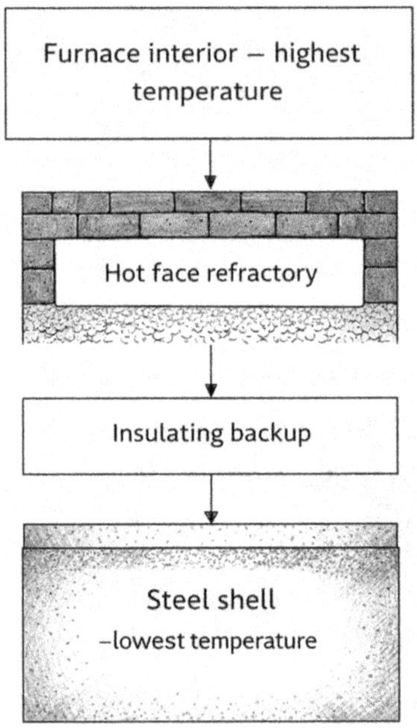

Figure 29.1 — Refractory Lining Cross Section
Typical layered structure of refractory linings inside high-temperature furnaces.

Industrial furnaces use layered refractory linings to manage extreme temperatures. The hot face refractory directly contacts the furnace interior and withstands the highest temperatures. Behind it, insulating backup materials reduce heat loss and protect the external steel shell, which remains at the lowest temperature to maintain structural stability.

Temperature does not remain uniform across thickness. It forms a gradient. The inner surface expands more than the outer region. This differential expansion creates internal stress. If the hot face is restrained by colder layers, tensile stress develops near the surface. Tensile stress is dangerous in brittle ceramics.

This is why expansion joints and compressible layers are not optional features. They are structural necessities.

When lining design ignores thermal gradient behaviour, cracking becomes inevitable.

Constraint and Anchoring

Castable refractories are often secured to steel shells using metallic anchors. These anchors experience the same heating but expand differently from ceramic material. Metal expands more. If anchors are rigidly embedded without allowance, stress concentrates at anchor tips.

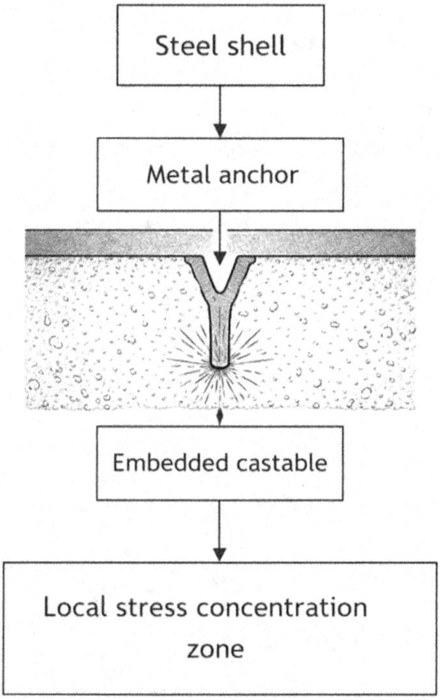

Figure 29.2 — Anchor System Layout
Anchors support refractory linings and accommodate thermal expansion in high-temperature installations.

In refractory linings, metal anchors are welded to the steel shell to support castable refractory materials. The anchor transfers mechanical loads and holds the refractory in position during thermal expansion and contraction. However, the anchor location can create a local stress concentration zone within the castable lining, which must be considered in refractory design.

Anchor-induced cracking often begins at hot face and propagates toward anchor root. In cyclic heating environments, repeated expansion and contraction amplify damage. Over time, cracks link and segments detach.

Correct anchor alloy selection, spacing, and geometry are critical. Constraint must be controlled, not eliminated, but distributed safely.

Refractory design is as much mechanical engineering as it is material science.

Refractoriness Under Load (RUL)

A refractory may not melt at operating temperature, yet it can still deform. Under sustained mechanical load, structural phases begin to soften.

RUL measures the temperature at which measurable deformation begins under constant load.

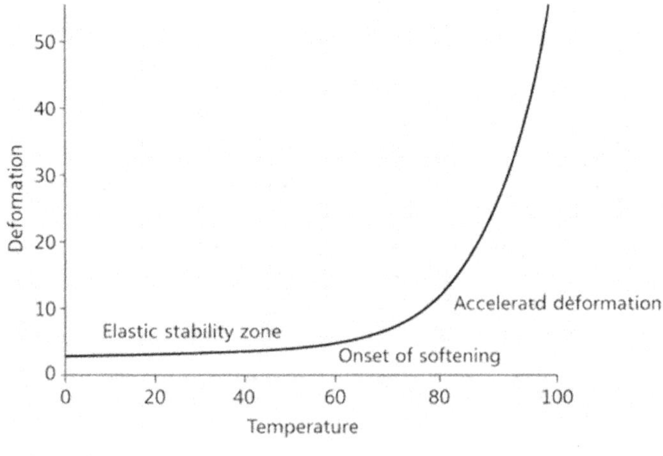

Figure 29.3 — RUL Curve

Curve showing deformation behaviour of refractories under load at increasing temperature.

As temperature increases during the refractoriness-under-load (RUL) test, the material initially remains stable in the elastic stability zone. With further temperature rise, softening begins and deformation starts to increase. Beyond this point, deformation accelerates rapidly as the refractory loses its load-bearing capability.

The RUL curve defines safe operating ceiling. If working temperature approaches the softening onset, structural integrity weakens. Even small deformation can open joints or distort alignment in rotary kilns.

Operating slightly below RUL limit provides stability. Operating near or above it shortens campaign life drastically.

Refractory selection must always include RUL margin, not only melting point.

Creep — Time as a Destructive Variable

Even below RUL temperature, time can cause deformation. This phenomenon is creep. Creep is slow, permanent strain under constant load and high temperature. It progresses in stages.

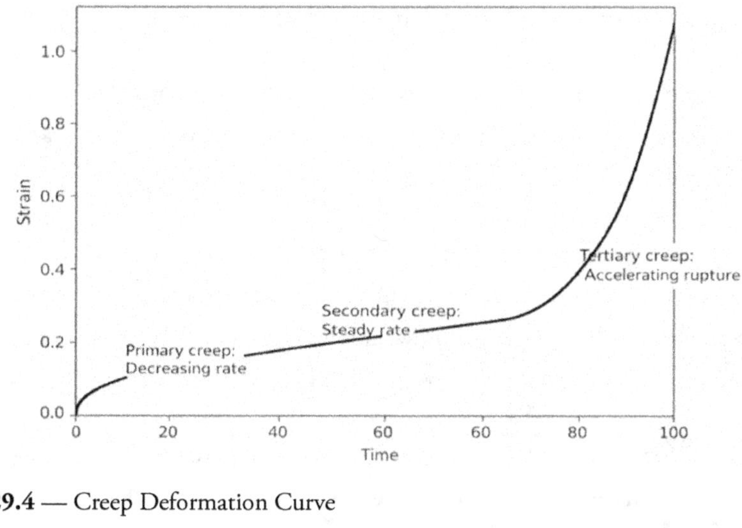

Figure 29.4 — Creep Deformation Curve

Illustration of time-dependent deformation of refractory materials at high temperature.

When a material is subjected to constant stress at elevated temperature, deformation progresses through three stages. Primary creep shows a decreasing deformation rate as the material adjusts to the applied load. Secondary creep follows with a nearly steady deformation rate. In tertiary creep, deformation accelerates rapidly due to internal damage until rupture occurs.

Primary creep stabilizes as internal structure rearranges. Secondary creep reflects steady microstructural deformation. Tertiary creep begins when internal damage accumulates and microcracks coalesce.

In furnace roofs or arches, creep can gradually reduce thickness. In vertical linings, it can cause bulging.

Grain size influences creep resistance. Coarse, well-bonded grains resist deformation better than fine, porous matrices. Glassy phases at high temperature increase creep susceptibility.

Creep is not visible early. But it determines service life.

Slag Attack — Chemical Degradation

Heat and load are only two forces. Chemistry adds the third.

Molten slag reacts with refractory oxides. Chemical compatibility determines resistance. Basic slag attacks silica-based refractories. Acidic slag attacks magnesia-based refractories.

When molten slag comes into contact with refractory materials, chemical reactions occur at the interface. The molten slag can then penetrate into the pore structure of the refractory. This penetration weakens grain boundaries and disrupts the microstructure, eventually leading to material loss and refractory degradation.

Slag dissolves protective phases. It penetrates open porosity. It reacts along grain boundaries. Mechanical flow removes reacted layer. Fresh surface is exposed. The cycle repeats.

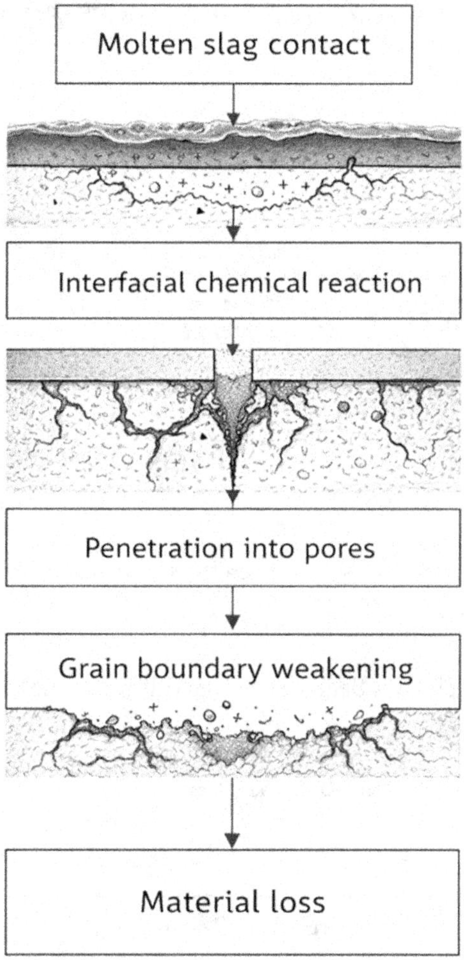

Figure 29.5 — Slag Attack Mechanism
Chemical interaction between molten slag and refractory material leading to structural degradation.

Dense microstructures resist penetration longer. Controlled porosity and compatible chemistry slow reaction.

Slag line zones experience maximum chemical stress. Most refractory failures initiate there.

Refractory Types and Application Matching

Material must match environment.

Table 29.1 — Refractory Types and Application Suitability

Type	Strengths	Typical Application
Fireclay	Economical, moderate resistance	General furnace lining
High Alumina	High temperature strength	Steel and cement zones
Magnesia	Basic slag resistance	Steel converters
Silica	Thermal stability at high temp	Coke ovens
Castables	Flexible installation	Complex geometries

No refractory is universally superior. Each is optimized for specific thermal and chemical conditions.

Wrong selection accelerates degradation regardless of thickness.

High-Temperature Failure Modes

Table 29.2 — High Temperature Failure Modes and Causes

Failure Mode	Structural Trigger
Thermal cracking	Rapid heating or cooling
Spalling	Repeated thermal cycling
Creep bulging	Sustained high load
Slag erosion	Chemical incompatibility
Anchor cracking	Expansion mismatch
Joint opening	Inadequate expansion allowance

Failure is rarely caused by a single variable. It is cumulative interaction.

Interaction of Heat, Load, and Time

Refractories operate under three simultaneous stresses: thermal stress from temperature gradients, mechanical stress from load and structural constraint, and chemical stress from slag or furnace atmosphere. Over time, each of these stresses intensifies.

Rapid heating increases thermal gradients, sustained load accelerates creep deformation, and chemical reactions weaken bonding within the structure. As these effects accumulate, microcracks begin to form, cracks link together, and fragments may eventually detach.

Refractory service life is therefore governed by balance. When temperature remains stable, creep tends to dominate. When heating and cooling cycles are frequent, thermal spalling becomes the primary damage mechanism. When the chemical environment is aggressive, erosion and chemical corrosion dominate.

Understanding which mechanism is dominant allows engineers to apply the correct corrective strategy.

Industrial Decision Insight

Refractory management is predictive, not reactive.

Before failure occurs, monitor:

Hot face temperature.
Shell temperature rise.
Thickness reduction at slag line.
Visible crack patterns.
Campaign duration trends.

Corrective actions include material upgrade, anchor redesign, controlled heating schedules, slag chemistry adjustment, or improved insulation.

Refractories fail progressively. Intelligent monitoring converts surprise failure into planned maintenance.

Structural Doctrine

Refractories are not fragile materials; they are highly engineered structural ceramics designed to operate where few materials can survive. However, they still obey fundamental physical principles. Thermal expansion must be accommodated, mechanical loads must remain within creep tolerance, chemical environments must be compatible, and heating must be properly controlled.

When these conditions are respected, refractory performance and service life remain stable. When they are ignored, degradation progresses gradually and often invisibly until failure appears sudden.

In the final part of this book, the discussion brings together all these mechanisms—surface defects, firing stress, raw material imbalance, and refractory degradation—into a unified failure intelligence framework. Ceramics do not fail randomly; they fail when the structure is pushed beyond its natural limits.

◈ Industrial Example 16 — Refractory Roof Sagging

Situation:
The furnace roof gradually sagged during a long operating campaign.

Initial Assumption:
The failure was attributed to weak or poorly designed anchors.

Reality:
Creep deformation of the refractory material at operating temperature exceeded the allowable structural limit.

Correction:
Use refractory materials with higher creep resistance suitable for sustained high-temperature loading.

Lesson:
Refractoriness Under Load (RUL) does not represent long-term creep resistance.

Explanation:

At high temperatures, refractory materials can slowly deform under sustained mechanical load through a process known as creep. Even when the applied stress is below the immediate failure limit, prolonged exposure to high temperature allows the material's microstructure to gradually deform.

Furnace roofs are particularly vulnerable because they must support their own weight over long periods. While RUL tests indicate the temperature at which deformation begins under load, they do not measure the rate of creep over extended service time. Selecting refractories with proven creep resistance ensures dimensional stability and prevents progressive sagging during long campaigns.

◈ Industrial Example 17 — Unexpected Refractory Spalling

Situation:

Spalling occurred in the refractory lining after repeated thermal cycling.

Initial Assumption:

The failure was attributed to poor refractory material quality.

Reality:

Thermal expansion mismatch between the hot-face refractory and the backup lining created stress during heating and cooling cycles.

Correction:

Select lining materials with compatible thermal expansion characteristics across all layers of the refractory system.

Lesson:

Systems fail, not materials alone.

Explanation:

Refractory linings often consist of multiple layers, each designed for a specific function such as thermal insulation, structural support, or corrosion resistance. When these materials expand and contract at different rates during heating and cooling, mechanical stresses develop at the interfaces between layers.

Repeated thermal cycling can cause these stresses to accumulate, eventually leading to cracking and spalling of the hot-face refractory. Proper refractory design requires considering the entire lining system, ensuring that materials are thermally compatible so that expansion differences do not generate damaging stresses.

❖ Industrial Example 18 — Thermal Cycling Damage in Cement Rotary Kiln Refractories

Situation:
Refractory linings in cement rotary kilns experienced surface spalling and gradual material loss during plant operation.

Initial assumption:
The refractory material contained manufacturing defects or insufficient mechanical strength.

Reality:
Repeated heating and cooling cycles during kiln start-up and shutdown created strong thermal gradients inside the refractory bricks.

Correction:
Control kiln heating schedules and select refractory materials with suitable thermal shock resistance for cyclic operating conditions.

Lesson:
Thermal cycling during kiln operation must be managed to prevent stress accumulation and refractory degradation.

Explanation:
During rapid heating or cooling, temperature differences develop between the hot surface and the cooler interior of the refractory brick. These

gradients create internal tensile stress. With repeated thermal cycles, small cracks gradually grow and cause surface layers to detach, a process known as spalling. Progressive material loss reduces lining thickness and increases heat transfer to the kiln shell. Controlled heating rates and appropriate refractory selection help reduce thermal gradients and extend lining service life.

Decision Insight:

- Slag chemistry must match refractory chemistry.
- RUL defines the safe operating temperature.
- Creep governs campaign life.

Common Mistakes in Ceramic Engineering

1. Increasing firing temperature to fix low strength.
   Often increases residual stress and deformation instead.

2. Blaming glaze for cracking without checking body expansion.

3. Assuming higher density always improves performance.

4. Ignoring cooling rate while focusing only on peak temperature.

5. Using thicker refractory instead of selecting correct chemistry.

6. Trusting cold MOR values for high-temperature applications.

7. Treating surface defects as cosmetic rather than structural.

8. Ignoring installation constraints when diagnosing field failure.

9. Over-deflocculating slip to increase casting rate.

10. Expecting laboratory results to directly match plant conditions.

These mistakes are not technical ignorance.

They are system misunderstandings.

Correct thinking prevents recurring failure.

Limits of Prediction in Ceramic Systems

Ceramics are governed by physical laws.
But they are also governed by statistics.

Strength depends on flaw size.
Flaw size distribution is never perfectly uniform.

Residual stresses are influenced by minor thermal variation.
Raw materials vary slightly from batch to batch.

Because ceramics are flaw-controlled materials, absolute prediction is impossible.

Engineering does not eliminate uncertainty.
It reduces it.

Testing reduces uncertainty.
Process control reduces uncertainty.
Material compatibility reduces uncertainty.

But no test can detect every flaw.
No inspection can see every microcrack.
No design can remove every stress concentration.

The goal of ceramic engineering is not certainty.
It is probability management.

When you understand structure, stress, and process history, you shift from reactive correction to predictive control.

Ceramics do not promise perfection.
They reward discipline.

Respect structure.
Reduce variability.
Design for margin.

That is the practical limit of prediction.

End of Part IX — Refractories

Extreme temperature does not forgive miscalculation.
Chemical attack, creep, and expansion mismatch operate slowly but relentlessly.
Material selection is a system decision.

Part X — Failure Intelligence

Diagnosing and Preventing Ceramic Failure

When ceramics fail, the visible defect is rarely the true origin of the problem. Failures develop through interactions between structure, stress, processing history, and service conditions.

This final section integrates the principles developed throughout the book and presents a structured approach to failure diagnosis and engineering decision-making.

Chapter 30

Why Ceramics Fail — The Unified Law of Behaviour

Failure is the visible result of accumulated imbalance in structure and stress.

Ceramic failure is never an accident. It is the visible end of an invisible sequence. When a ceramic component cracks, crazes, warps, spalls, or collapses, the event often appears sudden.

A tile fractures without warning. A refractory lining fails unexpectedly. A glaze develops cracks months after installation. To the observer, the event feels unpredictable.

But ceramics do not fail unpredictably. They fail when structure is pushed beyond tolerance.

Failure is not weakness. It is the consequence of law.

Throughout this book we examined raw materials, particle size distribution, plasticity, drying behaviour, sintering, glaze compatibility, refractory creep, slag attack, and surface defects. Each chapter revealed one part of CBS. The purpose of this final chapter is to integrate those parts into a single understanding.

Every ceramic failure, without exception, emerges from the interaction of stress, flaw, and constraint acting over time.

Failure is a Process, Not a Moment

A ceramic body begins as powder containing mineral phases, particle size distribution, impurities, and microscopic defects. During forming, density gradients may develop. During drying, shrinkage stresses arise. During firing, phase transformations alter the internal structure. During service, thermal and mechanical loads continue to act.

Failure may occur years after manufacturing, yet its origin often lies in earlier imbalance. The visible crack is rarely the beginning of the problem; it is the final stage of a long sequence.

A flaw introduced during forming may remain dormant. A thermal mismatch in glaze formulation may stay invisible. A refractory anchor stress concentration may appear harmless for months. But when additional stress aligns with the flaw and constraint prevents relief, fracture becomes unavoidable.

Ceramics accumulate structural imbalance until the tolerance threshold is eventually crossed.

Ceramic Failure Diagnosis Method (CFDM)

The Ceramic Failure Diagnosis Method (CFDM) provides a structured approach for identifying the root cause of ceramic failures.

The method follows the CBS by tracing failure backward through the stages of service conditions, properties, microstructure, processing, and raw materials.

This diagnostic approach allows engineers to identify the true origin of defects rather than only correcting surface symptoms.

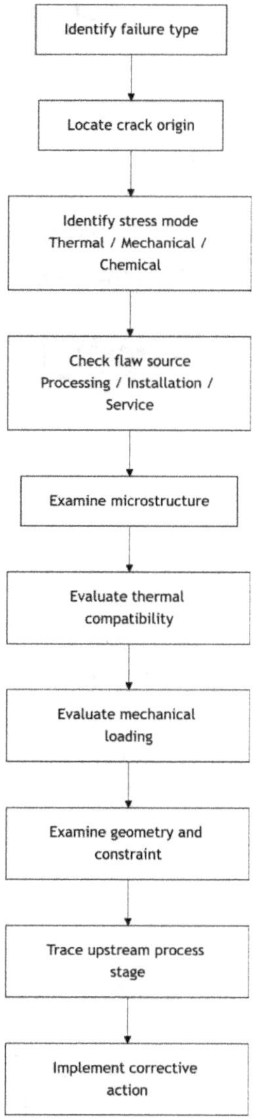

Figure 30.1 — Ceramic Failure Diagnosis Algorithm
Structured decision pathway used to trace the root cause of ceramic failure.

Failure diagnosis in ceramic systems follows a structured analytical sequence. The investigation begins by identifying the failure type and locating the crack origin.

The stress mode and possible flaw sources are then evaluated, followed by examination of microstructure, thermal compatibility, mechanical loading, and geometric constraints.

Finally, the upstream process stage responsible for the defect is traced, enabling the implementation of corrective actions.

The Diagnostic Discipline

Most industrial failure investigations fail because they focus on symptoms rather than mechanisms. Effective diagnosis requires structured discipline.

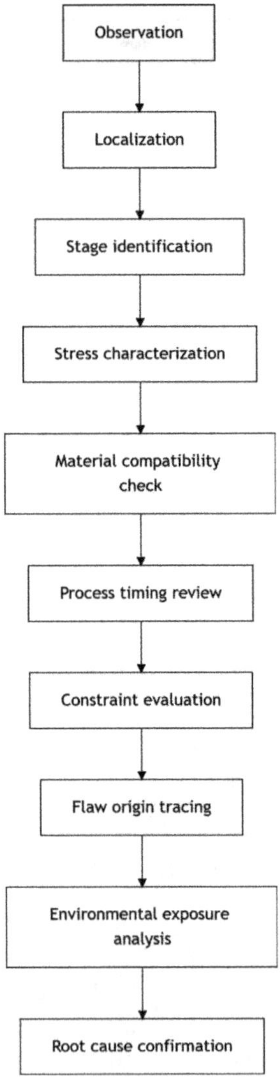

Figure 30.2 — Diagnostic Flowchart (10-Step Algorithm)
Step-by-step method for analysing ceramic defects and identifying failure origins.

Failure investigation in ceramic systems proceeds through a structured diagnostic flow. The process begins with observation and localization of the defect, followed by identification of the process stage and characterization of the stress involved.

Material compatibility, processing conditions, and structural constraints are then evaluated. By tracing the flaw origin and examining environmental exposure, the investigation ultimately leads to confirmation of the root cause.

This sequence forces the engineer to move from surface appearance to structural cause. Without structure, correction becomes guesswork.

For example, crazing appears as surface cracking. But the origin may lie in thermal expansion mismatch, firing curve variation, glaze thickness inconsistency, or moisture-assisted delayed crack growth. Only systematic elimination reveals the true mechanism.

Failure intelligence begins with disciplined questioning.

The Systems View of Failure

No ceramic component exists independently from its environment. Raw materials influence microstructure. Microstructure influences stress distribution. Stress interacts with constraint. Environment accelerates degradation.

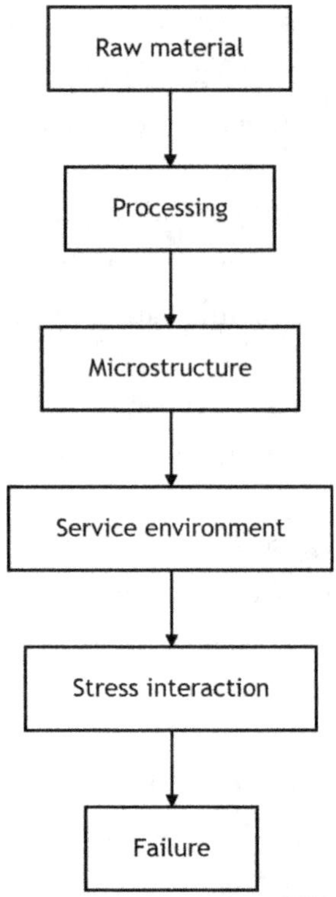

Figure 30.3 — Systems Failure Model
Ceramic failures result from the interaction of material, design, processing, and service conditions.

Ceramic failure is the result of a chain of interacting factors. Raw material characteristics influence processing behaviour, which determines the resulting microstructure. The microstructure then interacts with the service environment and applied stresses. Failure occurs when this combined system interaction exceeds the structural tolerance of the material.

This chain explains why correction at the final stage often fails.

Increasing firing temperature will not correct density gradient.
Increasing glaze thickness will not solve expansion mismatch.
Increasing refractory thickness will not prevent slag attack.

Failure is systemic. Systems thinking prevents repeated mistakes.

Stress in Ceramics

Ceramics are strong in compression and weak in tension. This asymmetry defines behaviour.

Thermal gradients create tensile stress at surfaces. Constraint converts expansion into tension. Cooling creates shrinkage tension. Mechanical load generates tensile stress at flaw tips.

Unlike metals, ceramics cannot plastically redistribute stress. They cannot yield to relieve concentration. Once stress intensity at a flaw tip exceeds fracture toughness, crack propagation occurs rapidly.

This explains why rapid heating cracks refractory surfaces, why fast cooling spalls kiln linings, and why glaze under tension forms crazing networks.

Stress must be controlled before it concentrates.

Flaws Are Inevitable

No ceramic material is completely free of defects. Pores, microcracks, grain boundaries, and inclusions are inherent.

Strength is therefore statistical. Larger flaws reduce strength. Wider flaw distributions increase failure probability.

Particle size distribution, forming uniformity, and firing control determine flaw size and distribution.

Failure occurs not because flaws exist — but because flaws become critical under applied stress.

Constraint — The Hidden Amplifier

Constraint is often underestimated in ceramic systems.

A glaze constrained by the body stores tensile stress.

A refractory constrained by steel shell stores expansion stress.

A tile constrained by rigid substrate stores installation stress.
Constraint transforms thermal expansion into tensile force.

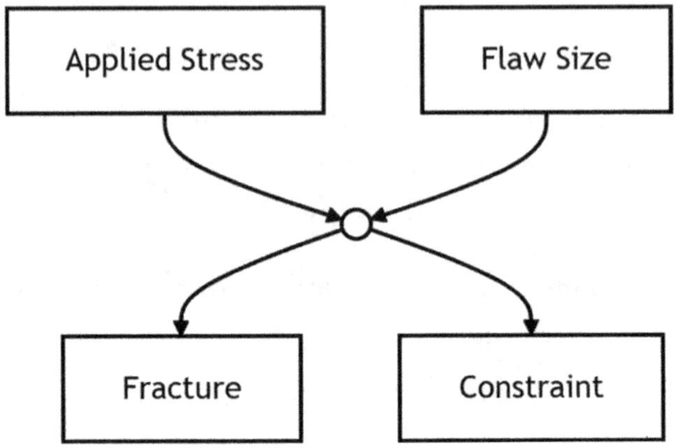

Figure 30.4 — Stress + Flaw + Constraint Model
Failure occurs when tensile stress acts on a flaw under constrained conditions.

A ceramic component fails when tensile stress interacts with a critical flaw under sufficient constraint. Reducing any one of these three variables can prevent fracture.

Remove any one variable and failure may not occur.

Reducing flaw size increases tolerance against stress.
Reducing constraint reduces stress intensity.
Reducing applied stress prevents crack propagation.

Failure prevention means adjusting at least one of these variables.

Economics of Failure

Technical failure always carries economic consequences.

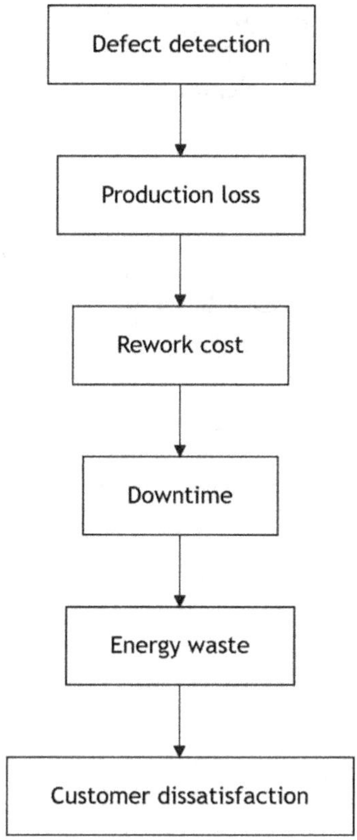

Figure 30.5 — Failure Economics Impact Chain
Ceramic failures propagate through production, maintenance, and operational costs.

A defect in ceramic production triggers a cascade of economic consequences. Detection of defects leads to production loss and additional rework costs.

These disruptions can cause equipment downtime and unnecessary energy consumption. Ultimately, persistent quality failures affect delivery reliability and lead to customer dissatisfaction.

A cracked refractory lining forces shutdown.
A glaze defect causes batch rejection.
A warped tile damages brand trust.

Failure intelligence is therefore not academic discipline. It is economic necessity.

Mapping Defects to Process

Understanding process origin accelerates root cause tracing.

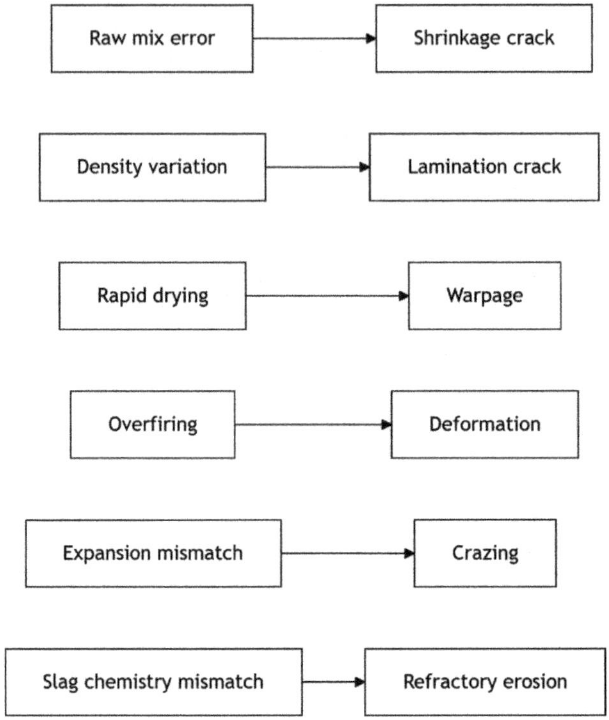

Figure 30.6 — Process–Defect Interaction Map
Mapping of manufacturing stages to typical defect formation mechanisms.

Defects in ceramic products often originate from specific process conditions or material interactions. Errors in raw mix composition can cause shrinkage cracks, while density variations during forming may lead to lamination cracks.

Rapid drying can produce warpage, and overfiring may cause deformation. Thermal expansion mismatch results in glaze crazing, whereas incompatibility between slag chemistry and refractory composition can lead to refractory erosion.

Instead of reacting emotionally to defects, engineers identify the process stage where imbalance began.

Industrial stability depends on process awareness.

The Ultimate Integrative Law

After studying every chapter, the final integration becomes clear.

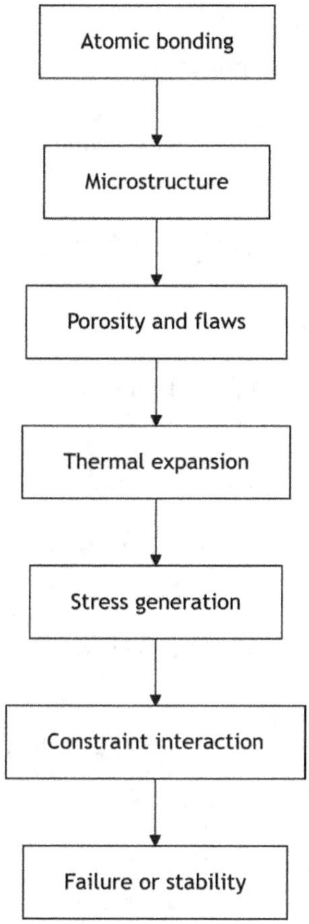

Figure 30.7 — Integrated Ceramic Failure Framework
Unified model linking materials, processing, structure, stress, and service environment.

Ceramic Behaviour System originates from atomic bonding, which governs the formation of microstructure during processing. The resulting microstructure contains porosity and inherent flaws that influence thermal expansion behaviour.

Thermal expansion generates internal stresses that interact with structural constraints. The combined interaction of these factors determines whether the ceramic system remains stable or ultimately fails.

Bonding defines brittleness.
Microstructure defines flaw distribution.
Porosity influences strength and permeability.
Thermal expansion determines stress magnitude.
Constraint determines stress amplification.

Ceramic stability exists only when these variables remain balanced.

Process Stage vs Defect Origin Mapping

Table 30.1 — Process Stage vs Defect Origin Mapping

Process Stage	Typical Defect	Structural Cause
Raw material preparation	Inclusions, contamination	Impurities or poor mixing
Forming / pressing	Lamination cracks	Air entrapment or uneven compaction
Drying	Drying cracks, warpage	Moisture gradient and shrinkage stress
Early firing	Black core, gas defects	Incomplete burnout
Sintering	Deformation	Uneven densification
Cooling	Thermal cracks	Rapid temperature change
Glazing	Crazing, crawling	CTE mismatch or contamination
Installation	Tile tenting, cracking	Restrained expansion
Service conditions	Spalling, fatigue cracking	Thermal cycling or load

Structured Diagnosis

Table 30.2 — 10-Step Diagnostic Framework Summary

Step	Core Question
1	What exactly failed?
2	Where did the crack initiate?
3	Which process stage introduced risk?
4	What stress dominated?
5	Was expansion compatibility maintained?
6	Was constraint excessive?
7	Were flaw sizes controlled?
8	Was chemical stability maintained?
9	Was timing properly controlled?
10	Which structural law was violated?

Failure Cost Impact Chain

Table 30.3 — Failure Cost Impact Chain

Failure Stage	Consequence	Cost Impact
Microstructural defect	Reduced strength margin	Hidden reliability risk
Early crack initiation	Local damage	Reduced component life
Crack propagation	Visible defect	Product rejection
Structural fracture	Component failure	Replacement cost
Operational disruption	Production interruption	Downtime losses
System damage	Secondary equipment damage	Repair cost
Safety risk	Personnel hazard	Liability risk
Reputation loss	Customer dissatisfaction	Long-term brand impact

The Unified Law of Ceramic Failure

Across all ceramic systems — tiles, sanitaryware, kiln furniture, refractories, and advanced ceramics — failure follows the same structural rule. Ceramic fracture occurs when tensile stress interacts with a critical flaw under constraint.

If any one of these three elements is absent, fracture may not occur. This principle explains most failures observed in ceramic engineering. A ceramic component may contain flaws yet remain stable if tensile stress is low. A component may experience stress yet remain stable if the flaw size is small. It may even experience both stress and flaws and still remain stable if expansion and movement are free and structural constraint is minimal. Failure becomes inevitable only when stress, flaw, and constraint combine beyond the structural tolerance of the material.

This rule applies across multiple levels of ceramic behaviour, including the microstructural level, the component level, and the system installation level. Recognizing this interaction transforms failure analysis from guesswork into engineering logic. Ceramic reliability therefore depends on controlling at least one of these three variables: reducing tensile stress, reducing flaw size, or reducing structural constraint. When these variables are managed properly, ceramic systems become stable and predictable. When they are ignored, failure becomes only a matter of time.

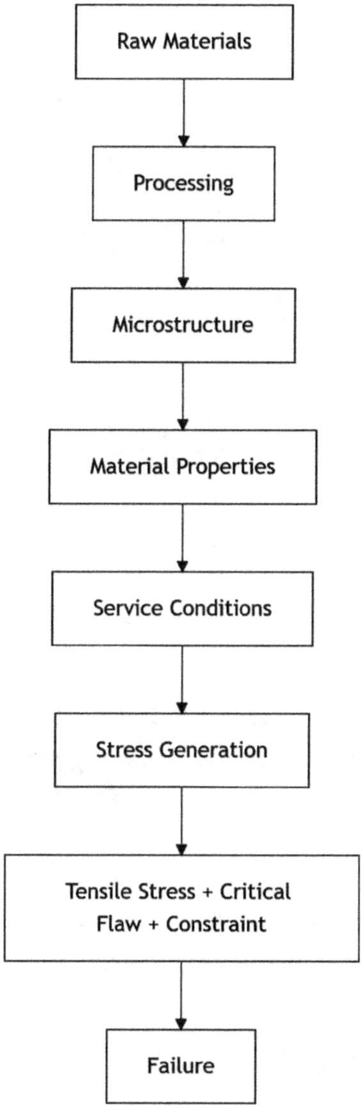

Figure 30.8 — CBS Failure Diagnosis Path
Failure analysis traced backward through the Ceramic Behaviour System chain.

Ceramic performance develops through a continuous sequence beginning with raw materials and processing. Processing establishes microstructure, microstructure determines properties, and properties interact with service conditions to generate stress. Failure occurs when tensile stress interacts with a critical flaw under constraint.

Figure 30.9 — Concept Map of Ceramic Behaviour
Overview diagram summarizing the relationships governing ceramic behaviour and failure.

Relationship between atomic structure, microstructure, flaws, and failure behaviour in ceramic materials, along with key influencing factors such as processing, temperature, environment, and loading.

Final Doctrine

Ceramics are powerful materials.
They resist heat, compression, and corrosion better than most structural systems.

But they demand respect.

They cannot absorb tensile stress casually.
They cannot redistribute stress through plastic deformation.
They cannot tolerate uncontrolled expansion mismatch.
They cannot survive aggressive chemistry without compatibility.

Failure is not the cruelty of the material. It is the consequence of ignoring structure.

If structure is understood, ceramics become predictable. If predictable, they become controllable. If controllable, they become reliable.

Reliability is not luck. It is intelligence applied to structure.

Ceramic engineering begins where guesswork ends.

Decision Insight:

- Trace failure upstream.
- Stress + flaw + constraint explains fracture.
- Respect structure before modifying process.

End of Part X — Failure Intelligence

Failure is not surprise.
It is accumulated imbalance.

When structure, stress, and environment align, ceramics perform.
When they do not, fracture reveals the truth.

Each part of this book builds on the previous one. As understanding accumulates, the Ceramic Behaviour System becomes clearer and failures become predictable.

How to Think About Ceramic Failures

Ceramic failures rarely occur where they first appear. A crack observed after firing may have originated much earlier—during forming, drying, or even raw material preparation. Ceramics are strong but sensitive to imbalance. Small differences in moisture, density, temperature, or constraint can create stresses that remain hidden until later stages of processing or service.

Because ceramics cannot deform significantly, they often store stress quietly before releasing it suddenly as cracks, warpage, or distortion. Experienced ceramic engineers therefore study the entire process history rather than focusing only on the stage where failure becomes visible.

When a defect appears, several questions must be considered: where could stress have been introduced, where could variation have formed, and where could constraint have prevented movement? Failure analysis begins with the observed defect and traces backward through the behaviour chain. Ceramic manufacturing becomes easier when behaviour is understood rather than when parameters are repeatedly adjusted. That is the purpose of this book.

Engineering Discipline

Engineering action is necessary, but overreaction can be dangerous. Not every defect requires a formulation change, not every crack demands a higher firing temperature, and not every refractory failure requires a thicker lining. Before modifying a process or material, it is important to pause and examine the situation carefully.

The key questions should always be asked first: Has the root cause been verified? Is the source of stress identified? Is the flaw origin confirmed? Is the failure isolated or systemic? Has variability been measured? Without answering these questions, corrective actions may address symptoms rather than the true cause.

For this reason, changes must be made with discipline. Do not increase firing temperature simply because measured strength appears low. Do not reduce cooling rate without confirming the origin of stress. Do not add flux to increase densification without checking the expansion balance of the body. Do not change glaze chemistry without measuring the body's coefficient of thermal expansion. Do not upgrade refractory grade without evaluating the chemical environment in which it operates.

Ceramic systems are interconnected, and one change can shift multiple variables simultaneously. The disciplined engineer understands when intervention is necessary, while the experienced engineer also knows when careful observation is the better response. In many cases stability is restored by correcting a small imbalance rather than introducing major changes. Before altering the structure of a ceramic system, it is essential to understand it fully.

The Mental Model You Now Carry

If this book has done its work, you no longer see crazing as random cracking, warpage as a drying accident, slag erosion as bad luck, refractory collapse as a surprise, or pinholes as a simple surface imperfection. Instead, you begin to see interaction.

You see stress seeking a flaw, expansion fighting constraint, timing controlling microstructure, and chemistry attacking structural weakness. This shift in perspective is the real outcome of study. Ceramic behaviour becomes understandable once these interactions are recognized.

Simplicity Is Structural Clarity

"Made Simple" does not mean reduced science. It means reduced confusion. Complex equations exist behind fracture mechanics, phase diagrams govern sintering behaviour, and thermodynamics explains slag reactions. Yet practical engineering intelligence does not require memorizing every equation. It requires understanding relationships.

Ceramic behaviour can be understood through simple structural relationships: stress interacting with flaws and constraint, heat interacting with time and chemistry, and structure interacting with environment and tolerance. Once these relationships are internalized, ceramic behaviour becomes readable rather than mysterious.

Final Integration

This book began with atomic bonding and ends with failure intelligence. Between those points lies the life cycle of every ceramic component:

Powder → Form → Dry → Fire → Use → Degrade → Fail

At every stage of this sequence, structure governs the outcome. If structure aligns with the service environment, ceramics survive. If structure exceeds its limits, ceramics fracture. There is no mystery beyond this structural relationship.

The Ten Structural Laws of Ceramics

The behaviour of ceramics can be summarized through ten structural principles.

1. Strength is controlled by the largest flaw.

2. Ceramics do not yield before fracture.

3. Expansion without freedom creates stress.

4. Cooling is as critical as heating.

5. Uniformity creates reliability.

6. Processing defines microstructure, and microstructure defines performance.

7. Cracks grow silently before failure.

8. Time reduces the safety margin.

9. Systems fail, not materials alone.

10. When structure is respected, failure becomes predictable.

These are not trends. They are structural realities. When these principles are understood, ceramics become clearer. When they are ignored, failure appears sudden even though it has been developing quietly within the structure.

These principles apply to:

- Tiles and sanitaryware
- Refractories
- Tableware
- Advanced ceramics
- Industrial Ceramic Products

These are time-independent principles.

Author's Note

This book was written to remove unnecessary fear from ceramics. Ceramic materials are often described as brittle and unpredictable, but in practice their behaviour follows clear structural principles linking microstructure, stress, temperature, and environment.

During many years of working with ceramic systems in industry, it became clear that most failures were not mysterious. They were the result of structural imbalance somewhere along the processing chain. The purpose of this book has therefore been not only to explain ceramic science, but to organize it into a framework that engineers can use in real decisions. Ceramics become simpler when their structure is understood.

Epilogue — The Law of Structure

Ceramics are often described as brittle and unpredictable. In reality, their behaviour follows a disciplined structure. From atomic bonding to kiln firing curves, from glaze fit to refractory creep, every behaviour follows structural relationships.

Strong bonds create stiffness. Limited slip creates brittleness. Porosity introduces flaws. Thermal expansion creates stress. Constraint amplifies tension. Chemistry accelerates degradation. Time accumulates damage.

The scale may change—tile, sanitaryware, refractories, glaze systems, or advanced ceramics—but the governing logic does not.

The Discipline of Respect

Ceramics reward discipline. Particle size must be controlled. Density uniformity must be maintained. Drying gradients must be managed. Heating rates must be controlled. Thermal expansion must be matched, movement must be allowed, and chemistry must be compatible with the service environment.

Each of these actions is an act of structural respect. When this discipline exists, performance stabilizes. When shortcuts appear, failure accelerates.

The Industrial Edge

Industries that understand ceramic structure outperform those that simply react to defects. Predictive maintenance replaces emergency shutdown. Process stability replaces batch rejection. Root cause correction replaces symptom patching. Ceramics reward systemic thinking.

The Closing Law

Ceramics are not fragile materials; they are precise materials. Precision is not weakness—it is discipline. When discipline guides design, processing, and operation, ceramics perform beyond expectation. When discipline is ignored, failure appears sudden, even though it has been developing quietly beneath the surface.

Understand the structure. Respect the structure.
And ceramics will not surprise you.

Appendices

Appendix A — Thermal Expansion Reference Table

Thermal Expansion Reference Table

Material Type	**Relative Thermal Expansion**
High quartz body	High
High alumina body	Moderate
Magnesia refractory	Moderate to high
Silica refractory	Low at high temp (after inversion)
Typical glaze (alkali rich)	Higher than body
Boron-rich glaze	Lower than alkali glaze

Note: Always verify CTE under actual firing conditions. Expansion compatibility must be evaluated as a system.

Appendix B — Failure Patterns

Common Ceramic Failure Patterns (Quick Recognition Guide)

Visible Pattern	**Likely Cause**
Network surface cracks	Glaze tensile stress
Edge flaking	Excess glaze compression
Lamination cracks	Density gradient during pressing
Corner cracking after firing	Rapid cooling
Warpage	Uneven shrinkage
Pinholes	Gas escape through glaze
Blisters	Trapped gas beneath glaze
Refractory hot-face erosion	Slag attack
Refractory roof sagging	Creep deformation

This table is intended for rapid industrial troubleshooting.

Appendix C — Process Checklist

Process Control Checklist

Before blaming material, verify:

Raw Materials

- Consistent particle size distribution
- Controlled moisture content
- No contamination

Forming

- Uniform pressing pressure
- No trapped air layers
- Controlled green density

Drying

- Gradual moisture removal
- Balanced airflow
- No surface-over-core gradient

Firing

- Controlled heating ramp
- Complete burnout before glaze sealing
- Uniform peak temperature

- Controlled cooling rate

Service

- Expansion allowances provided
- Compatible substrate
- Controlled environment exposure

Appendix D — Ceramic Discipline

The Ceramic Discipline

This book began with atomic bonding and ends with failure intelligence. Between these points lies the life cycle of every ceramic component:

powder → form → dry → fire → use → degrade → fail.

At every stage, structure governs the outcome. When structure aligns with the environment, ceramics survive; when structure exceeds its limits, ceramics fracture.

Ceramics are not fragile materials. They are precise materials. Precision is not weakness—it is discipline. Ceramic engineering therefore demands control: control of raw materials, control of processing, control of microstructure, and control of stress. When these elements remain balanced, ceramics perform reliably. When one element is neglected, imbalance accumulates until stress eventually reveals it.

Ceramics do not forgive assumptions; they reward structural understanding. Every crack tells a history, every failure reveals a sequence, and every successful component reflects discipline applied upstream. Ceramics are precise materials, and precision requires discipline.

**“Ceramics never fail without reason.
They only reveal the structure we created.”**

www.ingramcontent.com/pod-product-compliance
Lightning Source LLC
La Vergne TN
LVHW022200200926
848481LV00006B/1800
* 9 7 9 8 9 0 4 3 1 0 9 7 4 *